New Landscapes 景界 2

Commercial and Public Landscape

商业公共景观

佳图文化 编

华南理工大学出版社
·广州·

图书在版编目（CIP）数据

景界.2，商业公共景观 / 佳图文化编． —广州：华南理工大学出版社，2013.8
ISBN 978-7-5623-3950-2

Ⅰ．①景… Ⅱ．①佳… Ⅲ．①商业区-景观设计-作品集-世界 Ⅳ．① TU986.2

中国版本图书馆CIP数据核字（2013）第 119063 号

景界2：商业公共景观
佳图文化 编

出 版 人：	韩中伟
出版发行：	华南理工大学出版社
	（广州五山华南理工大学17号楼，邮编510640）
	http://www.scutpress.com.cn　E-mail: scutc13@scut.edu.cn
	营销部电话：020-87113487　87111048（传真）
策划编辑：	赖淑华
责任编辑：	李　欣　赖淑华
印 刷 者：	利丰雅高印刷（深圳）有限公司
开　　本：	1016mm×1370mm　1/16　印张：16.5
成品尺寸：	245mm×325mm
版　　次：	2013年8月第1版　2013年8月第1次印刷
定　　价：	296.00 元

版权所有　盗版必究　　印装差错　负责调换

Preface 前言

Commercial and public landscape plays an important role in urban public space, for it is the meeting point for varied commercial activities and it is closely connected with the urban master planning. Therefore, commercial and public landscape is created not only as the landscape but also shows a kind of social effect. Excellent landscape plan within commercial and public area will well combine the commercial spaces with the eco-environment, achieving environmental benefits and economic benefits at the same time.

In this collection, we've carefully selected the latest commercial and public landscape projects all over the world to analyze the planning and design for landscapes of this category. These projects are classified as cultural squares, urban spaces, office spaces, commercial spaces, healthcare and education spaces, etc. covering almost all kinds of commercial and public spaces. Each project is explained with keywords, features, design descriptions as well as professional drawings (plans, details, renderings, high-resolution photographs, etc.).

As a professional illustrated resource book on new landscapes, it includes the projects with excellent ideas and high technologies, representing the pioneering design for commercial and public landscapes. With rich content, exquisite composition and novel edition, we believe this new collection will bring you with the international ideas and global views.

商业公共空间在城市公共空间中占有重要地位，是现代城市各种商业活动的集聚地，它与城市总体规划设计的各个阶段都有着密切的关系。因而，商业公共景观不仅仅是一处景观，更是一种社会效益的体现。通过商业公共景观的合理规划，将生态环境与商业设计有机结合，达到城市环境效益与经济效益的和谐统一显得尤为重要。

本书以此为契机，精选世界各地极具代表性的最新商业公共景观案例，深入解读商业公共景观规划设计的始末。全书按照不同的景观类型，分为文化广场、城市空间、办公空间、商业空间、医疗教育空间等类别，基本囊括了各类商业公共景观。内容编排上，分别从景观案例的关键点、亮点、设计要点入手，配合大量的专业技术图纸，如平面图、细部图、效果图以及实景图等，资料丰富详实。

《景界2：商业公共景观》作为景观方面的最新专业读本，书中精选的案例可谓是设计灵感的迸发与技术完美结合的成果，代表了当下商业公共景观设计的前沿理念。通过丰富的内容、精心的编排以及新颖的展示，我们相信本书定会给景观设计师及相关行业读者带来世界级的先进理念和全球性的设计视野。

CONTENTS 目录

Cultural Square 文化广场

002	National 911 Memorial	美国911国家纪念园
008	Square Four	四角广场
014	Horno 3 Steel Museum	奥尔诺钢铁博物馆

Urban Space 城市空间

022	Public Space Design of Maria-Theresian-Straße in Innsbruck (IBK)	因斯布鲁克（IBK）Maria-Theresian-Straße 街道公共空间设计
028	The Avenue	宾夕法尼亚大道2200号
038	Haarlem Droste Site	哈勒姆市公共空间
044	Spectator Stand of Rowing Center in Bled	布莱德赛艇中心看台
048	"Potgieterstraat" Amsterdam	阿姆斯特丹 Potgieterstraat 儿童游乐街
054	France Beaucouzé City Center	法国 Beaucouzé 城市中心

Commercial Space 商业空间

060	Kettering Market Place	凯特林集市
064	200 5th Avenue New York City	纽约第五大街200号重建
070	Bodrum Primex Hotel	博德鲁姆 Primex 酒店
078	New Otani Hotel Roof Gardens	新大谷酒店屋顶花园
086	France Atoll Retail Park	法国 Atoll 零售公园
094	Crown Center Square	皇冠中心广场
100	Forum Homini Boutique Hotel	微型精品酒店景观
106	Poly International Plaza	保利国际广场

112   Gruau Industrial Park   Gruau 工业园

116   Cristalia Business Park   Cristalia 商业园

124   717 Bourke Street   717 Bourke 商业街

132   The "Grand" Hotel Amsterdam   阿姆斯特丹索菲特大酒店庭院

Office Space 办公空间

142   Sunnylands Center & Gardens in Rancho Mirage, CA   加州兰乔米拉奇阳光岛花园中心

150   Melbourne International Convention & Exhibition Center   墨尔本国际会展中心

158   California Academy of Sciences   加州科学馆

164   The Steel Yard   钢铁工厂院落

170   Gardens of Corporate Headquarters   公司总部花园

176   Gannett—USA Today Headquarters   Gannett——《今日美国》总部大楼

Healthcare and Education 医疗教育

186   Arizona State University Polytechnic Campus—New Academic Complex   亚利桑那州立大学理工学院——新教学综合大楼

194   Reitaku University—New School Building in the Woods   日本丽泽大学——树丛中的新教学楼

206   Campus Mall at Tohoku Pharmaceutical University   东北药科大学广场

216   Maastad Hospital, Rotterdam   鹿特丹玛莎塔德医院

224   Johannesburg City Park Environmental Education & Research Center   约翰内斯堡城市公园环境教育和研究中心

230   University of Johannesburg Arts Center   约翰内斯堡大学艺术中心

234   Manassas Park Elementary School Landscape   马纳萨斯公园小学景观

242   The Biodesign Institute at Arizona State University   亚利桑那州立大学生物设计研究所

248   Zambrano Hellion Hospital   Zambrano Hellion 医院

Cultural Square
文化广场

Cultural Identity
Spirit of Place
Sustainability
Creativity

文化属性
场所精神
可持续性
创意性

COMMERCIAL AND PUBLIC LANDSCAPE 商业公共景观

CULTURAL SQUARE 文化广场　　URBAN SPACE 城市空间

Keywords 关键词

- Solemn 肃穆
- Sustainable 可持续
- Eco-friendly 生态环保
- Evergreen Garden 常青园林

Location: New York, USA
Landscape Design: PWP Landscape Architecture

项目地点：美国纽约
景观设计：PWP景观公司

National 911 Memorial
美国 911 国家纪念园

Features 项目亮点

Through the integration of the infrastructure system, the design makes people cherishing the dead, mourning the sacred places with creativity and ecological harmony.

通过对基础设施系统的整合设计，使得人们缅怀死者、寄托哀思的神圣场所变得创意十足，并且构建生态和谐。

| COMMERCIAL SPACE 商业空间 | OFFICE SPACE 办公空间 | HEALTHCARE AND EDUCATION 医疗教育 |

Overview

Located in the center of the World Trade Center, National 911 Memorial covers about 8 acres of about 16 acres, as a part of the project to rebuild the World Trade Center. It is a sacred place to memory of the dead, of grief, and also a carefully create green space surrounding urban environment with a positive improvement. Memorial Park is composed of three main parts: the Memorial Pool, Memorial Square and Memorial Exhibition.

项目概况

美国911国家纪念园占地约8英亩（约32 375 m²），坐落在面积约16英亩（约64 750 m²）的世贸中心原址的中心，是重建世贸中心工程的一部分。它既是一处缅怀死者、寄托哀思的神圣场所，也是一个精心营造的绿色空间，其周边的环境也做出了积极的改善。纪念园由3个主要部分组成：纪念池、纪念广场及纪念展览馆。

| COMMERCIAL AND PUBLIC LANDSCAPE 商业公共景观 | CULTURAL SQUARE 文化广场 | URBAN SPACE 城市空间 |

Design Description

Forming a sound infrastructure system, the Memorial is similar to the concept of shared ditch, integrated irrigation, electricity, drainage pipelines. Making it easy to detect, adjust and repair to extend its service life.

The full thickness of the soil, irrigation, ventilation and drainage is the key to keep the woods green. Nearly 40,000 tons of planting soil have been shifted to the Memorial Plaza in order to guarantee that the oak can thrive.

The granite shop underground using sand cushion instead of cement mortar. When you need repairs, paving stones can easily be removed and may continue to use undamaged.

911 Memorial design creatively expressed reflection after the suffering and pain of the bereaved. The emotional design also embodies the philosophy of environmental protection and long-term perspective.

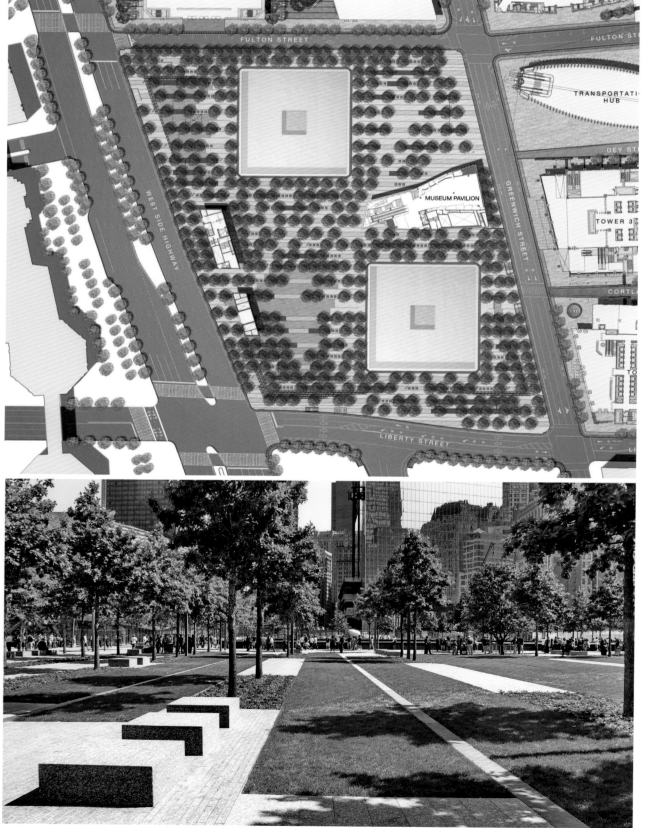

| COMMERCIAL SPACE 商业空间 | OFFICE SPACE 办公空间 | HEALTHCARE AND EDUCATION 医疗教育 |

设计说明

纪念园形成了一套完善的基础设施系统，它采用了沟渠共享、综合灌溉、电力集中供应、排水管线共用的概念，使之易于检测、调整和维修，以延长纪念园的使用寿命。

充分厚度的土壤，适当的灌溉、通气和排水，是纪念园林常青的关键。纪念广场移来了将近4万吨种植土，以保证橡树能茁壮生长。

地下花岗石商铺采用的是沙垫层而不是常用的水泥砂浆。当需要修缮时，这些铺地石块可以很轻松地被移开，并可无损坏地继续使用。

911纪念园的设计极具创意地表达了失去亲人的痛苦和痛后的反思，也体现了设计人的感性设计思想及长期的环保理念。

COMMERCIAL AND PUBLIC LANDSCAPE 商业公共景观

CULTURAL SQUARE 文化广场

URBAN SPACE 城市空间

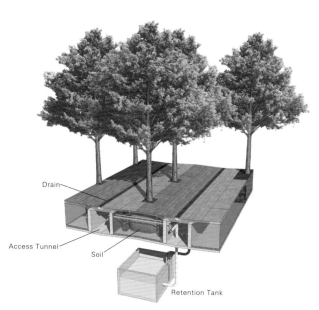

| COMMERCIAL SPACE 商业空间 | OFFICE SPACE 办公空间 | HEALTHCARE AND EDUCATION 医疗教育 |

| COMMERCIAL AND PUBLIC LANDSCAPE 商业公共景观 | CULTURAL SQUARE 文化广场 | URBAN SPACE 城市空间 |

Keywords 关键词

Venue Moods 场地精神
Quiet Space 静谧空间
City Imprint 城市印记
Visual Arts 视觉艺术

Location: Beirut, Lebanon
Landscape Design: Vladimir Djurovic Landscape Architecture
项目地点：黎巴嫩贝鲁特市
景观设计：Vladimir Djurovic 景观公司

Square Four
四角广场

Features 项目亮点

By grasping the venue moods, the design took the venue space, material substance, sound effects, light quality, even smell as the basic design elements to create a peaceful space.

设计较好地把握了场地精神，将场地的空间本身、物质实体、声响效果、光质乃至气味都视作基本的设计构成要素，创造出一个能让人远离尘嚣的空间。

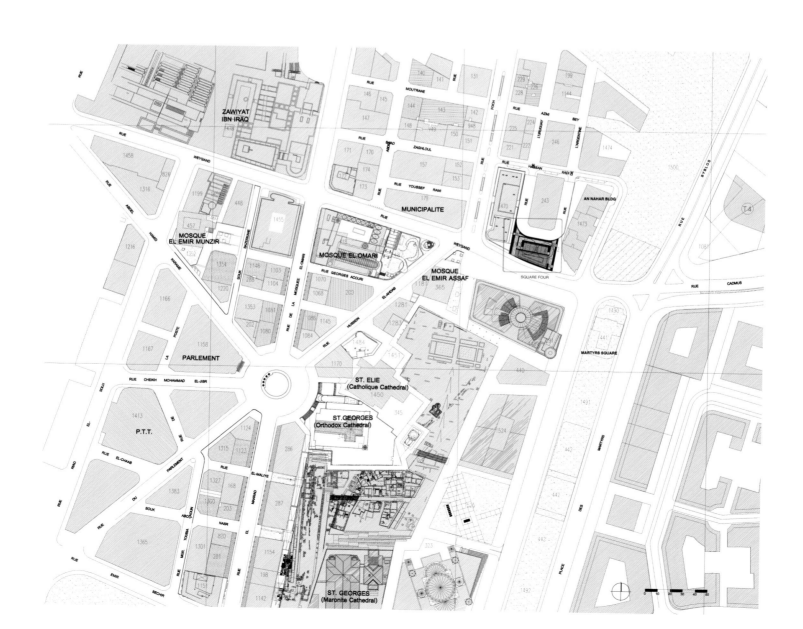

Overview

At the onset of Weygand Street, one of the main gateways to the newly redeveloped city center, between ancient ruins, new modern buildings, mosques and churches, Square Four emerges. Occupying a space of 815 square meters, it provides a haven of greenery, shade and calm within a busy urban setting.

项目概况

在通往新近重建的贝鲁特市中心的几条主干道之一的魏刚大街的起点,在古老的遗迹、现代的建筑,以及几座清真寺和教堂之间,四角广场脱颖而出。这个占地仅815 m² 的小广场,在繁闹的都市环境里创造了一个绿树浓荫的宁静港湾。

COMMERCIAL AND PUBLIC LANDSCAPE 商业公共景观

CULTURAL SQUARE 文化广场

URBAN SPACE 城市空间

Design Description

The square's design responds to an abstract conception of the area, where the space itself, materiality substance, sound effects, light quality, and even smells become its essential constituent elements. It is a space thought of as an escape from the city bustle while sited at the core of the city itself. A serene space that allows the city to speak about itself and its memory through the reflection of its skyline, with its mosques and churches, on the water surface, always sheltered by the two majestic ficus trees.

设计说明

广场的设计响应了空间的抽象理念，将场地的空间本身、物质实体、声响效果、光质乃至气味都视作基本的设计构成要素。在项目所处的市中心这个忙乱的环境中，设计力求创造出一个能让人远离尘嚣的空间。两棵高大榕树荫庇下的静谧小空间，通过清真寺、教堂和城市天际线投在水面上的倒影，向人们无声地诉说着贝鲁特这个城市以及关于它的历史记忆。

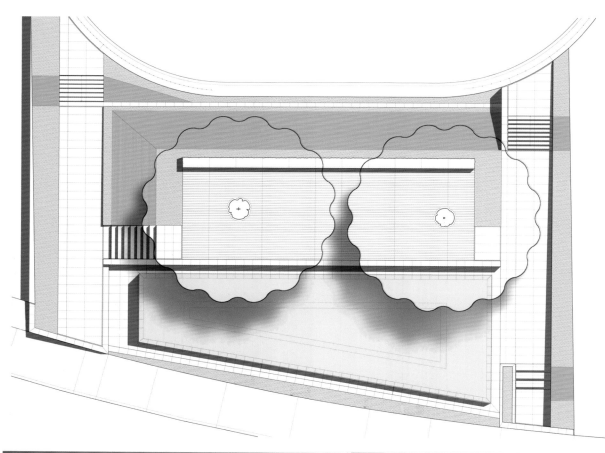

COMMERCIAL SPACE 商业空间	OFFICE SPACE 办公空间	HEALTHCARE AND EDUCATION 医疗教育

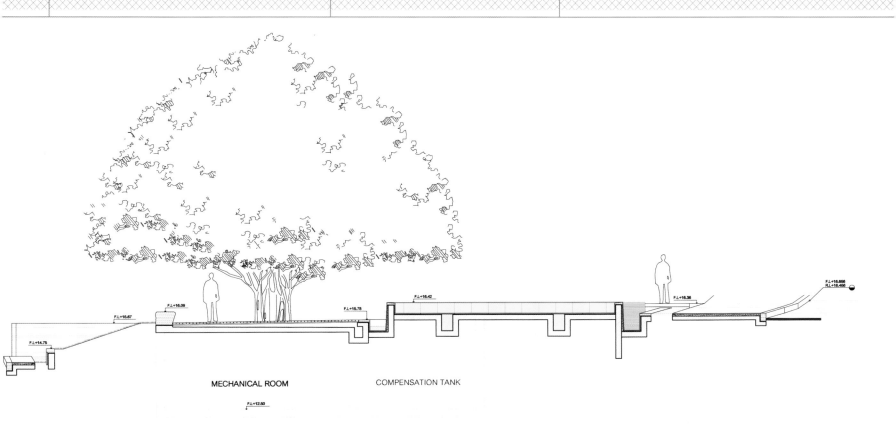

MECHANICAL ROOM   COMPENSATION TANK

COMMERCIAL AND PUBLIC LANDSCAPE 商业公共景观

CULTURAL SQUARE 文化广场

URBAN SPACE 城市空间

| COMMERCIAL SPACE 商业空间 | OFFICE SPACE 办公空间 | HEALTHCARE AND EDUCATION 医疗教育 |

COMMERCIAL AND PUBLIC LANDSCAPE 商业公共景观

CULTURAL SQUARE 文化广场　　URBAN SPACE 城市空间

Keywords 关键词

- Green Roof 绿色屋顶
- Landscape Terrace 观景露台
- Canal View 水道景观
- Sustainability 可持续性

Location: Fundidora Park, Monterrey, Nuevo León, México
Landscape Design: HARARI Landscape Architecture
Leader: Claudia Harari
Designers: Silverio Sierra, Diego Rodríguez, Lucía Narro, Paulina Cueva
Architectural Design: Grimshaw Architects, N.Y.

项目地点：墨西哥新莱昂州蒙特雷市高炉公园
景观设计：墨西哥HARARI景观设计事务所
项目负责人：Claudia Harari
设计团队：Silverio Sierra、Diego Rodríguez、Lucía Narro、Paulina Cueva
建筑设计：Grimshaw Architects, N.Y.

Horno 3 Steel Museum

奥尔诺钢铁博物馆

Features 项目亮点

Inspired by the industrial character of the site, the landscape uses mostly recycled materials (steel and concrete) and native grasses and sedums. It features the largest green roof built over a corten steel structure.

凭借遗址的特有资源，设计师充分利用回收的钢铁和混凝土，大量采用本地草坪和景天植物，建成了位于钢铁结构之上的最大的绿色屋顶。

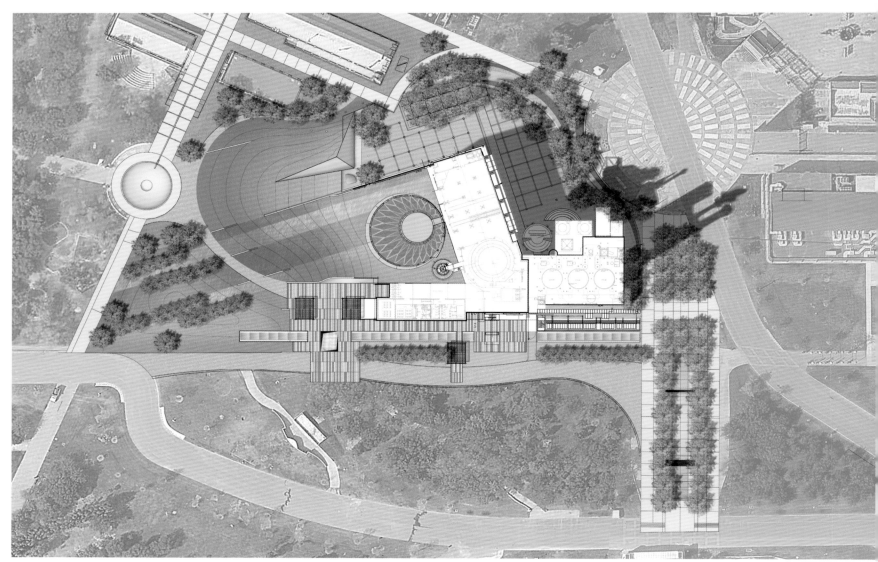

| COMMERCIAL SPACE 商业空间 | OFFICE SPACE 办公空间 | HEALTHCARE AND EDUCATION 医疗教育 |

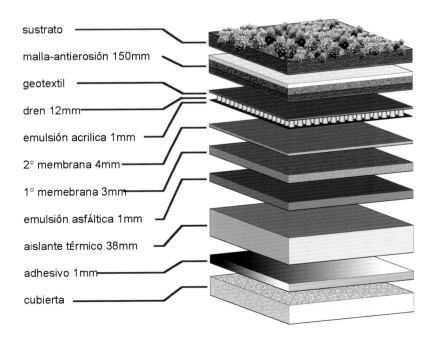

- sustrato
- malla-antierosión 150mm
- geotextil
- dren 12mm
- emulsión acrilica 1mm
- 2° membrana 4mm
- 1° memebrana 3mm
- emulsión asfáltica 1mm
- aislante térmico 38mm
- adhesivo 1mm
- cubierta

COMMERCIAL AND PUBLIC LANDSCAPE 商业公共景观

CULTURAL SQUARE 文化广场

URBAN SPACE 城市空间

Overview

A team of international designers collaborated to transform a former iron smelting furnace and an icon of the city into a modern history museum. Inspired by the industrial character of the site, the landscape uses mostly recycled materials (steel and concrete) and native grasses and sedums. It features the largest green roof built over a corten steel structure.

项目概况

一批国际设计师共同合作将一架退役的鼓风炉和一片废弃遗址改造成了这座现代历史博物馆。凭借遗址的特有资源，设计师充分利用回收的钢铁和混凝土，大量采用本地草坪和景天属植物，建成了位于钢铁结构之上的最大的绿色屋顶。

Design Description

The landscape for the Museo Del Acero Horno3 expresses the spirit of the site's former industrial glory and celebrates its position within the surrounding dramatic landscape. The overall landscape design emphasizes the physical profile of the 70-meter furnace structure while complementing the modern design of the new structures. The history of steel is an important narrative element throughout the site, and thus steel, much of it reclaimed from the site (such as the ore-embedded steel rails used to define the outdoor exhibit spaces) is used extensively to help define public plazas and delineate fountains and landscaped terraces. Large, free-formed steel objects and machinery unearthed during excavation were incorporated as stepping stones and other features. The design approach melds industrial site reclamation and the adaptive re-use of on-site materials with ecological restoration through the use of green technologies.

Two water features are integral to the narrative of the project, while helping to define and locate the public space adjacent to the museum. In the main esplanade, the steel plates that formerly clad the exterior of the main hall were repurposed into a stepped canal over which water

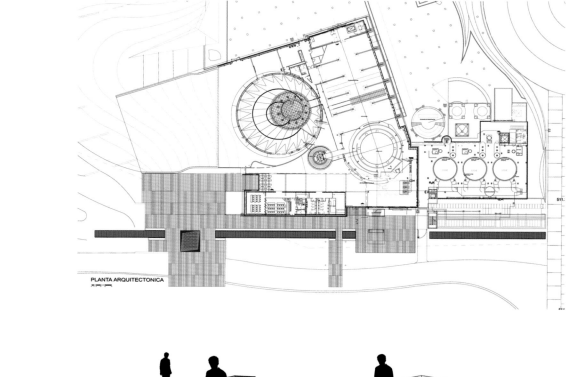

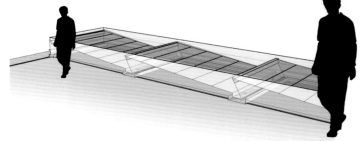

| COMMERCIAL SPACE 商业空间 | OFFICE SPACE 办公空间 | HEALTHCARE AND EDUCATION 医疗教育 |

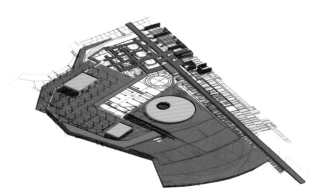

FOUNTAINS
OBJECTS/PLACES
PATHS
STRUCTURE
BORDERS/LANDFORMS

COMMERCIAL AND PUBLIC LANDSCAPE 商业公共景观 | CULTURAL SQUARE 文化广场 | URBAN SPACE 城市空间

cascades. This trompe l'oeil evokes the caustic heating process once used to extract ore, but instead of steam it generates a cooling mist that blows over the plaza – a pleasant surprise for visitors in Monterrey's hot and arid climate.

Principals of sustainability are at the core of the landscape design of the Museo Del Acero Horno3. By thoughtfully repurposing found industrial artifacts and incorporating new green technologies that work in concert with the architecture and the greater landscape, the designers have created an outdoor exhibition space that interprets the area's historic uses while celebrating artistic opportunities for the future.

This view of the alfombra verde (green blanket) shows the integration of the new museum with the surrounding park, while ameliorating the visual connection to the larger regional landscape.

The viewing terrace overlooks the sedum-planted roof and out toward the regional landscape beyond. The planting pattern works in concert with the structure of the new museum roof and evokes the flame patterns of the steel burner's former use

The landscape of the entrance to the Museo Del Acero Horno 3 evokes its industrial past and regional landscape while creating a sustainable public space for its future. The plaza is composed of recycled concrete, steel and artifacts while new green technologies create the museums new skin.

Set in a field of gravel, reclaimed ore-laden rocks comprise the misting water feature. This dramatic focal point evokes the historical steel manufacturing process while providing some welcome cooling effect in the arid climate.

As viewed from the retrofitted furnace elevator, a 200-meter reverse flow fountain creates the optical illusion of water flowing uphill. Extending from the museum entrance, the fountain's path represents the historical flow of materials in and out of the blast furnace.

In the evening, the steel in the canal water feature transforms into a cobalt blue glow. Lighting on the original furnace casts a dramatic silhouette on the site.

| COMMERCIAL SPACE 商业空间 | OFFICE SPACE 办公空间 | HEALTHCARE AND EDUCATION 医疗教育 |

设计说明

阿塞罗奥尔诺博物馆的景观让人们领略到它曾经辉煌的工业成就，它置身于周围这些引人注目的景观之中，亭亭玉立。总体景观设计突出了70 m高的鼓风炉的构造体型图并和新建筑的现代设计相得益彰。统观整座遗址，讲述钢铁的历史都显得尤为重要。因而，这些大部分是从遗址回收的钢铁（比如用于界定露天展区的含矿钢栏杆的钢铁）被广泛的用于界定公共休闲广场，装饰喷泉和景观露台。大块的形状不定的钢铁和挖掘遗址时未经发掘出来的机械装置混合改造成步石和其它一些特色景观。该设计方案通过使用绿色生态技术，使得工业遗址回收以及现场资源的选择性重复利用与生态修复完美地融合在了一起。

两处水的特色景观在该项目叙述中必不可少，它们同时也帮助界定和定位了博物馆邻近的公共空间。在主游憩区，以前用于主厅表面镀钢的钢材被重新用于建造有瀑布流过的阶梯式水道。这个视觉幻觉使人们联想起了开采矿石时的烧碱加热过程，但是不同的是，它并没有产生热腾腾的蒸气，取而代之的是缭绕在广场上方的凉爽的雾气。这无疑会给来气候炎热干燥的蒙特雷市旅行的游客带来不小的惊喜。

持久性原则是博物馆景观设计的核心。经过深思熟虑，设计者把回收的工业史前器物重新定位并结合新的绿色生态技术使其与建筑风格和更广阔的景观相得益彰，如此，他们创造出了这个露天展区。该展区诠释了这片土地具有历史意义的用途，同时也颂扬着它未来的艺术前程。

绿色地毯景观使得新博物馆和周围的公园成为整体，同时也改善了博物馆与更广域景观的视觉联系。

在观景露台上可以俯瞰植有景天属植物的屋顶和地区景观之外远处的风景。该种植模式不仅符合新博物馆屋顶的结构而且还能使人们联想到以前钢铁冶炼炉的火焰燃烧景象。

阿塞罗奥尔诺博物馆入口处的景观设计再现了它的工业历史和区域景观，同时也为其未来创造了持久性的公共空间。广场是由回收的混凝土，钢铁和工业史前器物建造而成，而新的生态技术也赋予了博物馆全新的面貌。

踏上一片砾石，含矿的再生岩石包含了薄雾水特色景观。这个引人入胜的焦点既给游客带来了酷暑中的一丝凉爽，也同时让人们联想起历史上钢铁的锻造过程。

从改装的鼓风炉升降机上看，会看到一条长200 m的倒流的喷泉，这使游客产生一种水往山上流的视觉假象。从博物馆入口处往外延伸，喷泉的路径也就是历史上出入鼓风炉的原料的流动路径。

水道景观中的钢铁在夜晚闪耀着美丽的钴蓝色辉光。灯光洒落到原始鼓风炉上，勾勒出了它美艳动人的轮廓。

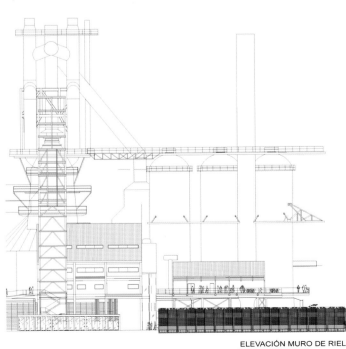

Urban Space
城市空间

Leisurely Atmosphere
Publicity
Lighting Design
Pavement

闲适氛围

公共性

照明设计

铺装材质

COMMERCIAL AND PUBLIC LANDSCAPE 商业公共景观

CULTURAL SQUARE 文化广场　　URBAN SPACE 城市空间

Keywords 关键词
- Leisure 休闲
- Dualities 双重性
- Materials 材料
- Public Space 公共空间

Location: Innsbruck, Austria
Client: Stadt Innsbruck
Planning: AllesWirdGut Architektur ZT GmbH
Area: 7,500 m²
Photography: Hertha Hurnaus

项目地点：奥地利因斯布鲁克
客　　户：Stadt Innsbruck
规划设计：奥地利 AllesWirdGut Architektur ZT GmbH
面　　积：7 500 m²
摄　　影：Hertha Hurnaus

Public Space Design of Maria-Theresian-Straße in Innsbruck (IBK)

因斯布鲁克（IBK）Maria-Theresian-Straße 街道公共空间设计

Features 项目亮点

Two defining materials, granite and brass balance the tense of land; different kinds of granite slabs with brass colour create a public space with relaxed and leisure atmosphere.

运用花岗岩和黄铜这两种材料平衡用地关系，不同类型的平面花岗岩铺装以及黄铜色调，烘托出这一公共空间轻松休闲的氛围。

| COMMERCIAL SPACE 商业空间 | OFFICE SPACE 办公空间 | HEALTHCARE AND EDUCATION 医疗教育 |

| COMMERCIAL AND PUBLIC LANDSCAPE 商业公共景观 | CULTURAL SQUARE 文化广场 | URBAN SPACE 城市空间 |

Overview

After redesigning of Maria-Theresian-Straße in the townscape of Innsbruck, the goal was to create an urban site with a rich atmosphere that invites strolling, hanging out, and meeting people.

项目概况

因斯布鲁克小镇上的 Maria-Theresian-Straße 经过重新设计改造，打造出氛围浓郁的、供人们外出散步和会面的休闲城市场所。

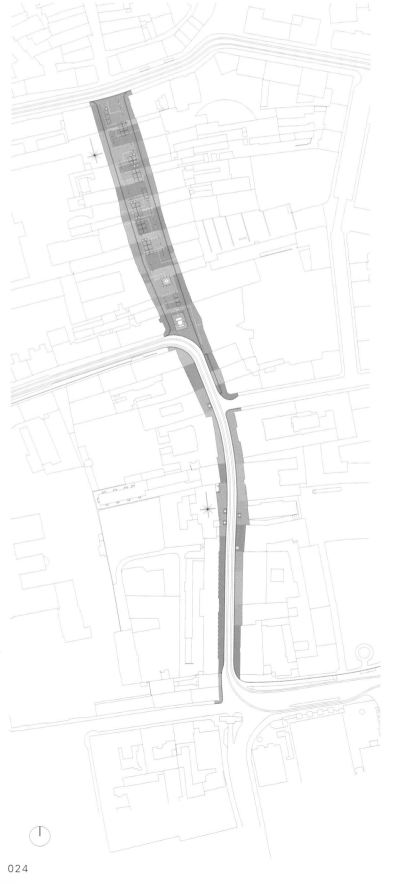

| COMMERCIAL SPACE 商业空间 | OFFICE SPACE 办公空间 | HEALTHCARE AND EDUCATION 医疗教育 |

Design Description

The identity of the site derives from the tension between urbanity and a panoramic view into nature, between past and future, between a specific character and a connective function in the urban structure of Innsbruck. Two defining materials, granite and brass balance these dualities in the redesigning: a slab carpet of four different types of granite creates a coherent square surface, and a network of brass-colored ground plates with street furniture growing up from it defines the square area proper in the middle of the street. At night, the walking zones alongside the house façades are brightly lit, while low-set lighting in the middle of the square enables a view of the mountain silhouette and the stars above.

设计说明

项目所在地所处的紧张关系来自都市化和自然全景之间，过去和未来之间，因斯布鲁克城市结构的特点和连接功能之间。这种双重性在重新设计中通过花岗岩和黄铜两种材料得到了平衡。四种不同类型的平面花岗岩铺设出一个连贯的广场表面，黄铜色的接地导板和街道家具在街道中间恰当地勾勒出广场的区域。夜晚，沿着住宅区立面的步行区灯火通明，广场中央低位设置的照明灯具创造出山峦的轮廓和天上的星星。

COMMERCIAL AND PUBLIC LANDSCAPE 商业公共景观

CULTURAL SQUARE 文化广场

URBAN SPACE 城市空间

| COMMERCIAL SPACE 商业空间 | OFFICE SPACE 办公空间 | HEALTHCARE AND EDUCATION 医疗教育 |

COMMERCIAL AND PUBLIC LANDSCAPE 商业公共景观

CULTURAL SQUARE 文化广场　　URBAN SPACE 城市空间

Keywords 关键词
- Open Space 开放空间
- Mixed-use 混合用途
- Waterscape 水景
- Stormwater Management System 雨水管理系统

Location: Washington, D.C. USA
Client: Boston Properties, Inc.
Design: SASAKI Associates
Area: 14,214 square meters - full site
6,628 square meters - streetscape & courtyards
(48% of site area)

项目地点：美国华盛顿特区
客　　户：波士顿地产
设　　计：美国SASAKI事务所
面　　积：14 214 m²
街道景观 & 庭院面积：6 828 m²（占面积的48%）

The Avenue
宾夕法尼亚大道 2200 号

Features 项目亮点

The waterscape in the courtyard functions as part of the large stormwater management system that has a multiple-effect on ecological and energy saving and practicality.

在中央庭院设置一处水景，作为雨水管理系统的一部分，达到生态、节能、实用的多重效果。

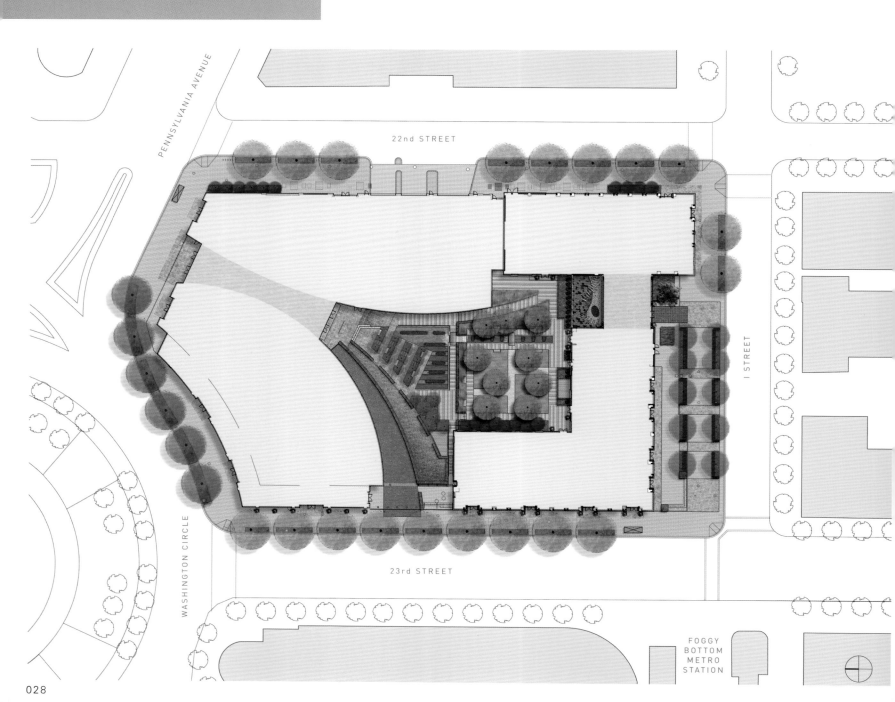

| COMMERCIAL SPACE 商业空间 | OFFICE SPACE 办公空间 | HEALTHCARE AND EDUCATION 医疗教育 |

Overview

The 2200 Pennsylvania Avenue, formerly referred to as Square 54, is a dynamic mixed-use development bordered by Washington Circle. Located on 23rd Street and Pennsylvania Avenue and just six blocks northwest of the White House. Also near George Washington University and close to a major public transportation hub, the entire-block complex includes office, residential and retail elements and abundant green public spaces, streetscapes, terraces, and courtyards with innovative stormwater management strategies implemented throughout. These spaces afford visitors, office building employees, and residents a pleasurable outdoor experience in all seasons.

项目概况

宾夕法尼亚大道2200号以前称为54广场，是华盛顿一个充满活力的混合用途发展圈，位于华盛顿23街、宾夕法尼亚大道，与东南方向的白宫仅隔六个街区。乔治华盛顿大学和主要交通枢纽近在咫尺。整个综合体包括办公、住宅、零售空间、丰富的绿化空间、街景、露台和全面配有先进雨水管理系统的庭院。一年四季为来访者、办公室工作人员和住户提供了一个令人愉悦的户外空间。

COMMERCIAL AND PUBLIC LANDSCAPE 商业公共景观

CULTURAL SQUARE 文化广场

URBAN SPACE 城市空间

| COMMERCIAL SPACE 商业空间 | OFFICE SPACE 办公空间 | HEALTHCARE AND EDUCATION 医疗教育 |

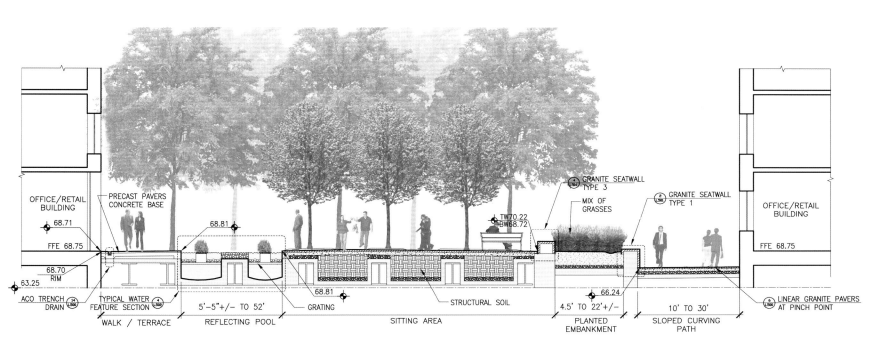

COMMERCIAL AND PUBLIC LANDSCAPE 商业公共景观

CULTURAL SQUARE 文化广场

URBAN SPACE 城市空间

| COMMERCIAL SPACE 商业空间 | OFFICE SPACE 办公空间 | HEALTHCARE AND EDUCATION 医疗教育 |

COMMERCIAL AND PUBLIC LANDSCAPE 商业公共景观

CULTURAL SQUARE 文化广场

URBAN SPACE 城市空间

Design Description

The footprints of the four buildings at Square 54 are designed to promote public use of the open space within the complex. The surrounding streetscape includes wide sidewalk promenades, bordered by rows of shade trees, large planting beds filled with mixed perennials, low shrubs and flowering trees, and a series of architectural planters filled with colorful seasonal plantings. All parking is located below grade within a five-story parking garage beneath the development. The central courtyard above the parking structure is anchored by a water feature that expresses the intersection of the historic Washington city grid and the axis of Pennsylvania Avenue. This water feature functions as part of the large stormwater management system that collects all rainwater that falls within the property. The water then drains through a stormwater filter to a 7,500 gallon cistern located in the five-story parking garage below the courtyard. This water is continuously re-circulated and treated by the water feature that includes aquatic plantings which offer supplemental filtration. The stored water is also used to provide all irrigation for the courtyard plantings throughout the growing seasons. The roof of the development contains 8,000 square feet of extensive green roof, which forms a microclimate that reduces the local heat island effect, provides avian habitat, insulates the building, and minimizes the roof's runoff. Excess rainwater is filtered through the green roof layers before being collected in the water feature and cistern below.

| COMMERCIAL SPACE 商业空间 | OFFICE SPACE 办公空间 | HEALTHCARE AND EDUCATION 医疗教育 |

设计说明

设计旨在将开放空间的公共作用最大化。周围的街景包括宽阔的人行漫步道、排排绿荫、充满着混合多年生植物的大花坛、低矮的灌木、开花的树木和五颜六色的当季植物。五层停车场位于综合体的地下空间。停车场上方的中央庭院有一处水景，正是历史悠久的华盛顿城市网络与宾夕法尼亚大道轴线相交的地方。水景也作为雨水管理系统的一部分，收集该综合体内的雨水。雨水通过过滤器排到位于停车场中的7 500加仑的水箱中。水景（包括水生植物）对雨水不断进行分流与处理以达到再次过滤的目的。收集的雨水可以用于所有植物的灌溉。屋顶有8 000平方英尺（约743 m²）的绿地，形成小气候，降低了局部的热岛效应，提供鸟类栖息地，避免建筑遭到暴晒，并最大限度地减少屋顶径流。多余的雨水在这里过滤再流入水景和水箱中。

COMMERCIAL AND PUBLIC LANDSCAPE 商业公共景观

CULTURAL SQUARE 文化广场

URBAN SPACE 城市空间

| COMMERCIAL SPACE 商业空间 | OFFICE SPACE 办公空间 | HEALTHCARE AND EDUCATION 医疗教育 |

037

COMMERCIAL AND PUBLIC LANDSCAPE 商业公共景观

CULTURAL SQUARE 文化广场　　URBAN SPACE 城市空间

Keywords 关键词
- Site Renovation 场地改造
- Riverfront Landscape 滨河景观
- Promenade 公共步道
- Lighting 特色照明

Location: Haarlem, The Netherlands
Client: DMV Vastgoed
　　　　IMCA Vastgoed
Landscape Design: Hosper Landscape Architecture
Designers: Mark van Rijnberk, Marike Oudijk, Jonas Strous
Area: 22,000 m²
Photography: Pieter Kers

项目地点：荷兰哈勒姆市
客　　户：DMV Vastgoed
　　　　　IMCA Vastgoed
景观设计：Hosper 景观设计事务所
设计团队：Mark van Rijnberk、Marike Oudijk、Jonas Strous
面　　积：22 000 m²
摄　　影：Pieter Kers

Haarlem Droste Site
哈勒姆市公共空间

Features 项目亮点

The plan combines the industrial past with modern urban needs, and makes the public spaces the linking elements for the area.

该城市空间在规划与设计时将工业历史与公共诉求有机结合，改建后的公共空间成为了联系所有区域的纽带。

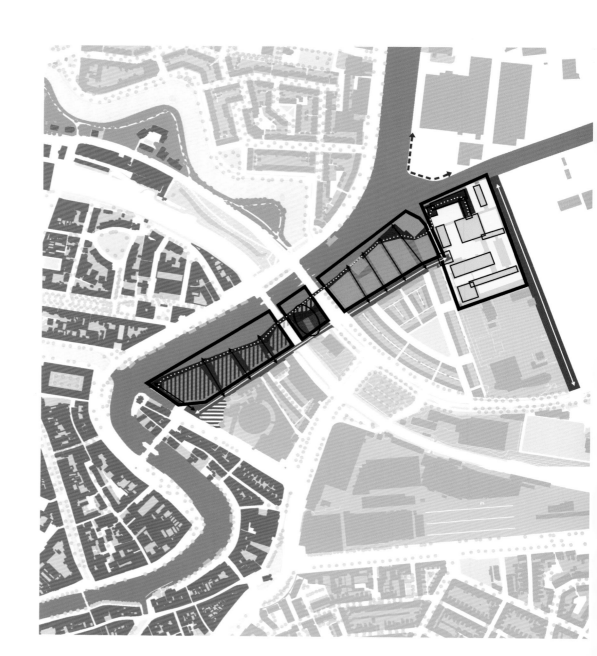

| COMMERCIAL SPACE 商业空间 | OFFICE SPACE 办公空间 | HEALTHCARE AND EDUCATION 医疗教育 |

Overview

During the coming years a number of obsolete factory sites along the River Spaarne in the centre of Haarlem will be redeveloped for homes, employment, culture and entertainment. The location that stirs the imagination mostly may well be the site of former Droste chocolate factory: it is in a fantastic location, has robust industrial monuments and a rich past.

项目概况

在未来几年内，哈勒姆市中心沿着 Spaarne 河岸边孤立的工业区将被改建为住宅、写字楼、文化及娱乐区域。其中最为显著的区域要属原 Droste 巧克力工厂所在地：该场地地理位置特别，具有深厚的工业背景。

| COMMERCIAL AND PUBLIC LANDSCAPE 商业公共景观 | CULTURAL SQUARE 文化广场 | URBAN SPACE 城市空间 |

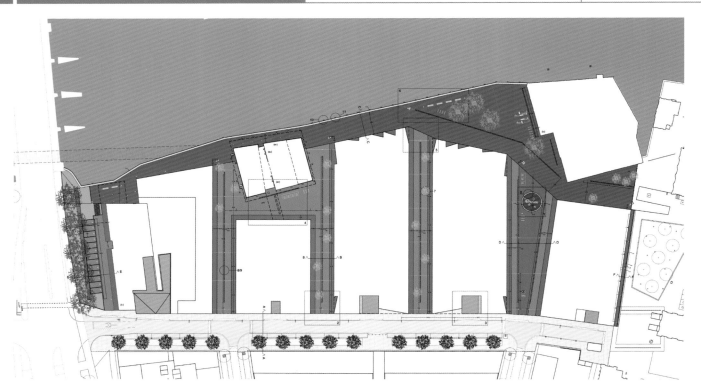
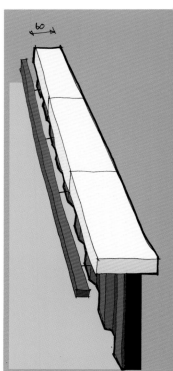

| COMMERCIAL SPACE 商业空间 | OFFICE SPACE 办公空间 | HEALTHCARE AND EDUCATION 医疗教育 |

Design Description

DKV has drawn up an urban plan for the site. This plan is based on the restoration of the monumental buildings, reuse of the later factory and construction of a number of new blocks of buildings. A broad promenade along the bank of the River Spaarne will end at a catering establishment in the monument. A number of pedestrian residential streets will be built at right angles to the bank zone. The public spaces are designed to be the linking elements for the area. The design is determined by the location along the River Spaarne, the site's industrial past, and the restricted car access to the region. The paving is comprised of one type of characteristic slab. A small size of rough, pitch-black copper slag paving slabs with a grit surface was selected. The pavements immediately in front of the homes are paved with contrasting granite cobblestones. Fresh-green, fragile ginkgos are planted at various locations. A group of trees of heaven with black crowns will be planted in front of the monument. The lighting is attached to the façades or suspended from overhead cables whenever possible. Over-sized masts with floodlights will be installed along the quay. A concrete capping piece with a buffer beam and mooring rings designed especially for the quay will enable pleasure boats to moor alongside. Some simple playground equipment will be grouped together in the broad street, the wedge, which ends the monument. The playground equipment will stand on a jet-black rubber surface in the form of the characteristic Droste logo. In addition to this playful reminder of the past, the factory buildings have also been combed for elements that can be reused in the development. This has resulted in a plan for the use of elements of the industrial heritage including tile tableaus, factory machines, and electrical switchgear cabinets.

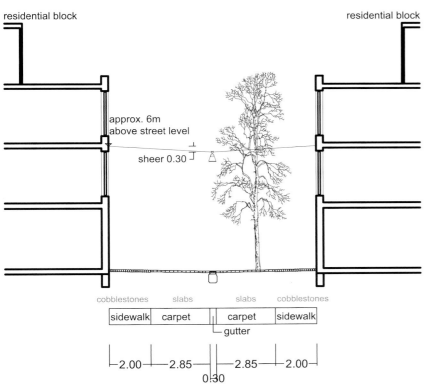

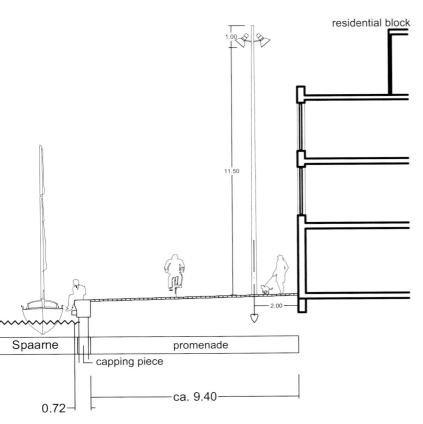

COMMERCIAL AND PUBLIC LANDSCAPE 商业公共景观

CULTURAL SQUARE 文化广场

URBAN SPACE 城市空间

| COMMERCIAL SPACE 商业空间 | OFFICE SPACE 办公空间 | HEALTHCARE AND EDUCATION 医疗教育 |

设计说明

KV 为该区草拟了城市规划方案。此次城市规划以具有纪念意义建筑的翻新、工厂等建筑的重新利用和新建筑的建造为基础。沿着 Spaarne 河岸有一条宽阔的散步道通向路尽头的一家餐饮公司。设计师们还将新建一系列通往河岸的道路。该地区的公共空间是连接所有区域的纽带。本设计主要是根据该区域沿 Spaarne 河岸的地理位置，该区域的工业历史以及通过该区域的限制车辆。公共地区的路面是由粗糙的小尺寸黑铜板铺成的，住宅前面的道路则是由鹅卵石铺成。市内不同的地区种植银杏，而纪念碑前面将种植黑冠臭椿。区域内照明系统齐全，码头边的专属设施也独具特色，还安装了简单的娱乐设施。照明沿外墙设计或尽可能地使用架空电缆悬挂起来。巨大的照明灯柱沿着码头布置。带有缓冲梁的混凝土顶盖及闸门设计的系泊环将为游艇停泊码头提供极大便利。一些简单的游乐设备将被组合在一起，置于宽阔的街道上。游乐设备将摆放在带有 Droste 标志的漆黑橡胶铺装上。除了这个联系过往的元素，旧厂房也被重新梳理成为新项目的组成部分。由此，一项合理利用工业遗产的方案被提出，包括利用旧的瓷砖、工厂的机器及电气开关柜等。

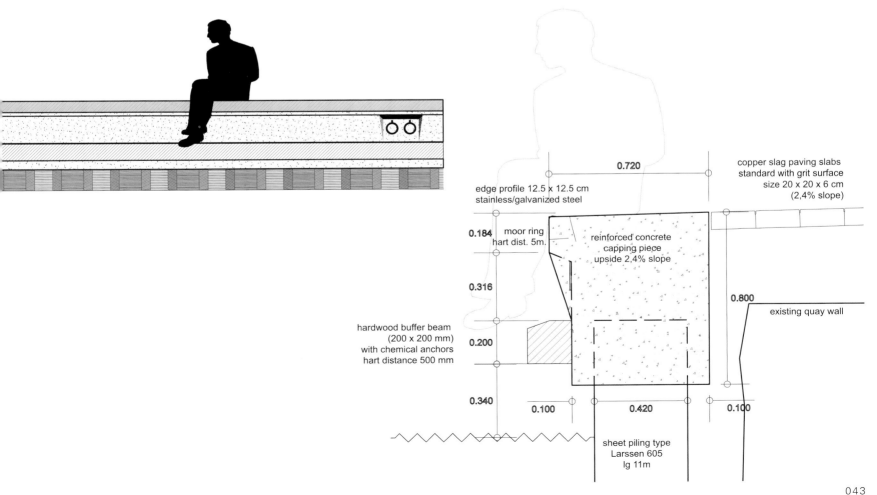

| COMMERCIAL AND PUBLIC LANDSCAPE 商业公共景观 | CULTURAL SQUARE 文化广场 | URBAN SPACE 城市空间 |

Keywords 关键词

Roof Terrace 屋顶露台
Open 开放性
Space 空间
Vision 视觉

Location: Velika Zaka, Bled, Slovenija
Client: Mestna Občina Bled
Architects: multiPlan arhitekti d.o.o.
Site Area: 712 m²
Built-up Area: 462 m²
Net Area: 520 m²

项目地点：斯洛文尼亚布莱德 Velika Zaka
客　　户：Mestna Občina Bled
建筑设计：斯洛文尼亚 multiPlan arhitekti 建筑师事务所
占地面积：712 m²
建筑面积：462 m²
净面积：520 m²

Spectator Stand of Rowing Center in Bled

布莱德赛艇中心看台

Features 项目亮点

As a spetactor stand for a sports game, the project not only pays attention to the noninterference between the vision line and the space considering its functional requirements, but also holds the site features to build a unique urban landscape architecture.

作为一个体育比赛的看台，本案设计既满足了其功能性的要求，注重视线与空间不受干扰，同时又很好地把握了场地特征，塑造了一个独特的城市景观建筑。

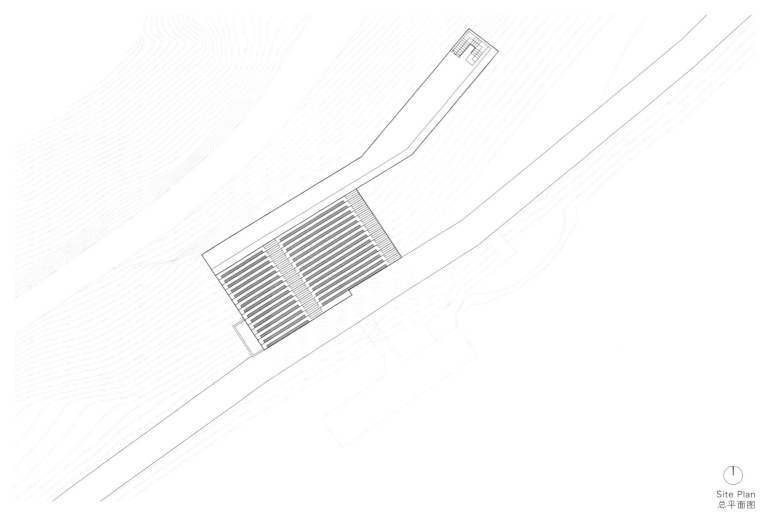

Site Plan
总平面图

| COMMERCIAL SPACE 商业空间 | OFFICE SPACE 办公空间 | HEALTHCARE AND EDUCATION 医疗教育 |

Overview

Bled is considered one of the most important and beautiful scenes of the largest rowing events. In 2011, it hosted the World Rowing Championships, where spectators cheered for the rowers from new stands in the finish line area. The renovation of the rowing center was funded by the European Regional Development Fund.

项目概况

布莱德被认为是举办大型赛艇比赛的最重要和最美丽的场地之一。2011年世界赛艇锦标赛曾在这里举行，看台位于比赛终点区域，观众坐在新建的赛艇中心看台为到达的运动员加油助威。赛艇中心的改造与维修是由欧盟地区发展基金赞助的。

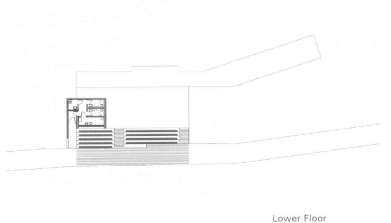

Lower Floor
下层楼

COMMERCIAL AND PUBLIC LANDSCAPE 商业公共景观

CULTURAL SQUARE 文化广场

URBAN SPACE 城市空间

Design Description

The stand facility is defined by two sets: the lower "open" viewing platform with seats and a closed, uniform area, designated to commentators and as VIP room. Above the object is a walked-on roof terrace, providing an exclusive view and capturing the whole vast image of the town with the lake and the island in a typical unique silhouette of Bled. The building is distinguished by a rational use of space under the stands, where storage facilities and toilets for visitors are placed. This part of the building is almost invisibly accessed from the stands, and there is no visual and spatial interference with the rowing promenade. The spatial intervention in the shore is thus limited to a minimum. The minimal level of intervention is supplemented by a canvas roof, reaching from the covered part of the stands and erected only during summer events.

设计说明

看台分为两个部分：下方配有座椅的开放式看台和一个统一的封闭区域，为评论员和VIP指定席。拾级而上是一个屋顶露台，在这里人们可以欣赏到独特的景色，一览整个城市的风景以及布莱德湖与岛屿的独特剪影。项目建筑的特点表现为空间的合理使用，在不显眼处为游客提供存储设施和厕所，可以通过看台直达，不会给赛艇带来任何视觉和空间的干扰。湖岸边的空间干扰被降至最小，这最小化的干扰也被帆布顶棚所弥补，只在夏季才打开遮挡住看台的一部分。

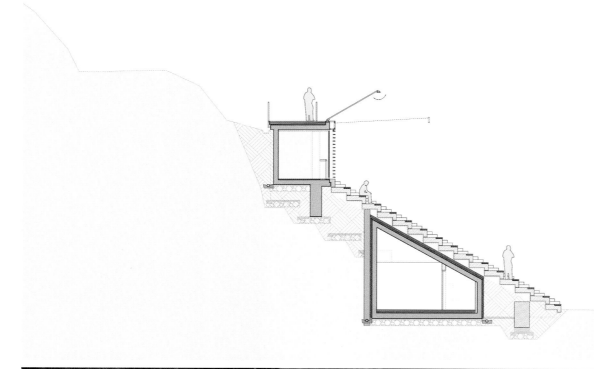

Sectional Drawing
剖面图

| COMMERCIAL SPACE 商业空间 | OFFICE SPACE 办公空间 | HEALTHCARE AND EDUCATION 医疗教育 |

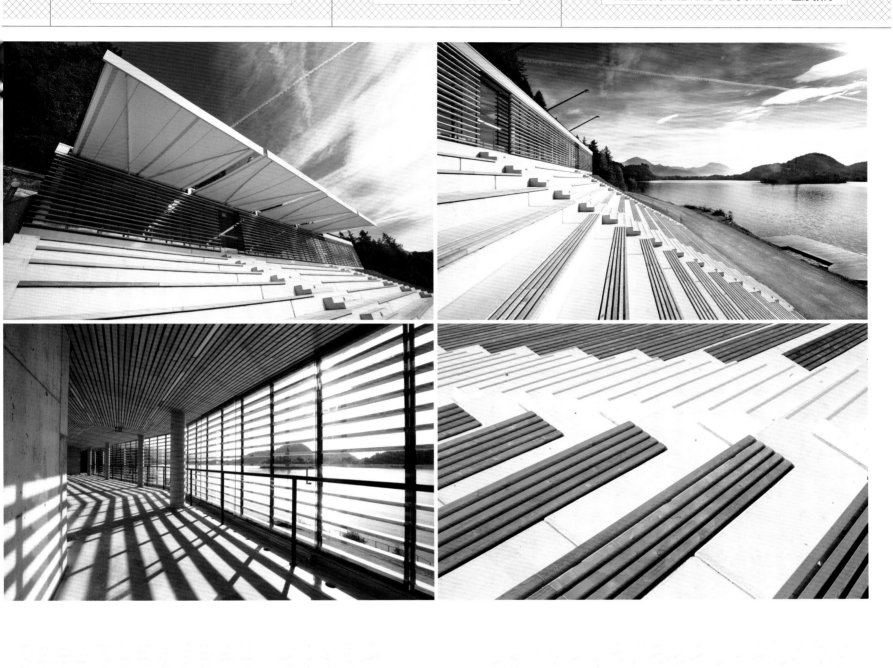

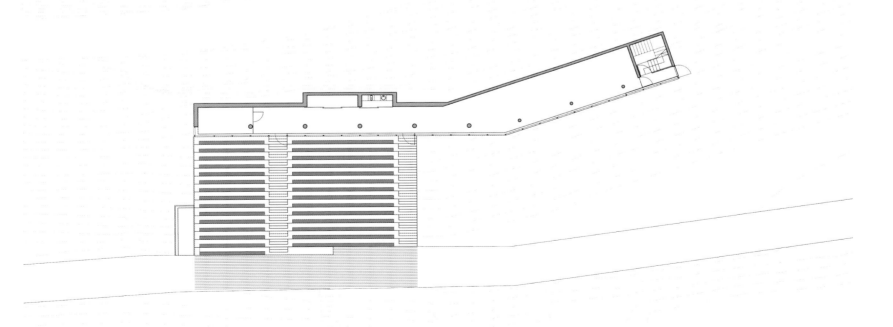

Ground Floor Plan
地面层平面图

COMMERCIAL AND PUBLIC LANDSCAPE 商业公共景观

CULTURAL SQUARE 文化广场   URBAN SPACE 城市空间

Keywords 关键词

Street Space 街道空间
Public Green 公共绿化
Children's Playground 儿童乐园
Neighborhood 社区环境

Location: Amsterdam, The Netherlands
Design Team: Elger Blitz, Mark van der Eng, Renet Korthals Altes, Jasper van der Schaaf, Lucas Beukers, Stef van Campen
Architects: Carve
Area: 1,500 m²

项目地点：荷兰阿姆斯特丹
项目设计：Carve
设计团队：Elger Blitz, Mark van der Eng, Renet Korthals Altes, Jasper van der Schaaf, Lucas Beukers, Stef van Campen
面　积：1 500 m²

"Potgieterstraat" Amsterdam

阿姆斯特丹 Potgieterstraat 儿童游乐街

Features 项目亮点

Local involvement in the design for this street became the stage for public participation. Old spaces are combined and turned to be an active street square. Playing facilities will enable the local children to enjoy themselves.

街道的设计具有广泛的群众参与性，原本封闭的街区空间被设计成一个合理有序的活力社区，而娱乐设施的配置让当地的孩子有了更广阔的玩耍空间。

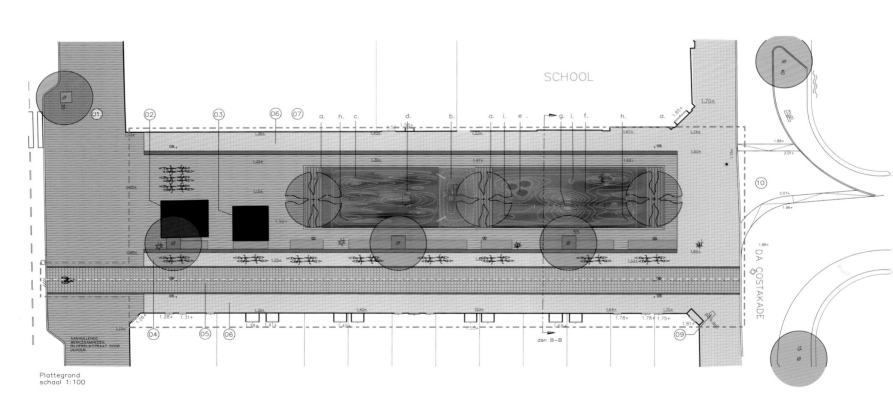

| COMMERCIAL SPACE 商业空间 | OFFICE SPACE 办公空间 | HEALTHCARE AND EDUCATION 医疗教育 |

Overview

Small interventions often evoke bigger changes. Local involvement in the design for the street in the city of Amsterdam became the stage for public participation.

项目概况

小举动通常会带来大变化。在阿姆斯特丹市,这项全民参与的街道设计中,公众积极投入,共同打造了这条充满乐趣的街道。

COMMERCIAL AND PUBLIC LANDSCAPE 商业公共景观

CULTURAL SQUARE 文化广场

URBAN SPACE 城市空间

Background

Potgieterstraat is situated in inner Amsterdam, in a context of 19th century buildings dating back to the first big enlargement of Amsterdam. The block typology of that time appears to the disadvantage of todays public life, since the inner courtyards of these blocks are not open to public use and the streets were never designed for today's traffic. In general there is a lack of public squares and public green. Streets here are dominated by cars and recently introduced bike lanes are a traffic solution, unfortunately claiming the available space from adjacent side walks. The district as a whole was up to a refreshing new strategy for children and pedestrians to strengthen and vitalize the public realm. Local inhabitants were asked in a political enquiry to agree upon and formulate new guidelines and were also involved in the selection of an architect.

设计背景

Potgieterstraat位于阿姆斯特丹市内，周围是19世纪阿姆斯特丹首次大扩张时期的建筑。彼时的住区环境已经不能适应当今社会的公众生活，因为建筑的内部庭院是封闭的，而街道也不能满足现在的交通需求。总的来说，该区缺少公共广场和公共绿化。目前的街道被汽车占据，现在虽设置了自行车道，但是又占据了临近的人行走道。这里急需进行重新整合，为孩子和行人腾出空间，并激活该地的社区活力。政府广泛征询当地居民的意见，鼓励人们对新规划出谋划策，并亲自参与设计师的评选。

| COMMERCIAL SPACE 商业空间 | OFFICE SPACE 办公空间 | HEALTHCARE AND EDUCATION 医疗教育 |

Aim of Intervention

Carve suggested to close down one of the streets entirely for car traffic in order to rededicate that street space to citizen utilization. The rededicated former street and parking area given to Carve to design, has a total site surface of 1,500 m². The sites functional program was changed from traffic and parking into an urban program of meeting and pausing places, a playground for children, an upgrading of the green quality and overall, a positive urban beacon for the district.

设计目标

面对城市公共空间不足的问题，Carve 建议将一条行车道路和泊车空间改建成占地 1 500 m² 的公共休憩场地，只有行人和单车可进入。在街道两边，除加强植树及绿化，更将其中一段街道打造成儿童游乐场，将游戏空间及社区设施的建构方式重新定义。

| COMMERCIAL AND PUBLIC LANDSCAPE 商业公共景观 | CULTURAL SQUARE 文化广场 | URBAN SPACE 城市空间 |

Design Description

Carve's intervention was firstly to rethink the street into a play street, accessible only to bikes and pedestrians. All surface materials were removed, the existing trees however were kept and new ones added. Into that clearance, Carve designed a mogul landscape with play objects integrated, materialized in abstract black rubber. The play objects vary from interactive elements to water sprayers. The rubber can be used as a drawing surface, invites to jump, run, fall thanks to its soft feel while reducing noise levels.However, the true benefit of this design is not obvious on a first glimpse. It is rather the reclaiming of local urban realm by its community. Parents but also citizens without children interact and relax here on wooden benches and around a little kiosk. The location becomes an anchor for neighborhood interaction and interlocks as well its surrounding blocks as well as helping to get together people of different backgrounds and ages. Residents even organize for instance little outdoor dinners and drinks here.

设计说明

改造工程由设计单位 Carve 负责，他首先考虑将此改造为仅供行人和单车进入的休憩地，把街道表面以前的障碍扫除，在保留原有的树木之外，也种植新的树木。设计师用黑色的橡胶铺砌成"波涛"的地面，上面有滑梯、管道和弹床，小朋友可随处涂鸦画画和攀爬跳跃。软软的橡胶不但加强保护，也有减低噪音作用。游乐场为小孩提供玩乐空间之余，也成为邻近社区的聚集点，居民可以在木椅上小坐或在小亭子里休息，甚至可以在这里举办小型的露天晚餐，促进了社区融和。

| COMMERCIAL SPACE 商业空间 | OFFICE SPACE 办公空间 | HEALTHCARE AND EDUCATION 医疗教育 |

COMMERCIAL AND PUBLIC LANDSCAPE 商业公共景观

CULTURAL SQUARE 文化广场　　URBAN SPACE 城市空间

Keywords 关键词
Landscape Furniture 景观小品
City Space 城市空间
Leisure Footpath 休闲步道
Lawn 草坪

Location: Beaucouzé, France
Client: Town of Beaucouzé
Project Management: Atelier Paul Arène
Area: 3,600 m²

项目地点：法国 Beaucouzé
客　　户：Beaucouzé 市政
项目管理：法国 Paul Arène 工作室
面　　积：3 600 m²

France Beaucouzé City Center
法国 Beaucouzé 城市中心

Features 项目亮点

The project offers wide and open spaces. The plants, landscape furniture and architectural layout are all for a comfortable environment.

景观设计所要体现的是一种低调而含蓄的开放性城市空间，所有的植被、景观小品以及建筑布局等均是为了营造休闲、舒适的生活环境而服务的。

| COMMERCIAL SPACE 商业空间 | OFFICE SPACE 办公空间 | HEALTHCARE AND EDUCATION 医疗教育 |

Overview

Landscape development aims to bring together different places of the everyday life around a common area, the surroundings of the townhall.

项目概况

景观开发的目标是将日常生活所需的各种休闲之地集合，围绕市政厅展开。

COMMERCIAL AND PUBLIC LANDSCAPE 商业公共景观

| CULTURAL SQUARE 文化广场 | URBAN SPACE 城市空间 |

| COMMERCIAL SPACE 商业空间 | OFFICE SPACE 办公空间 | HEALTHCARE AND EDUCATION 医疗教育 |

Design Description

Understated, like the pure architecture lines of the building, the project offers wide and open spaces. Planes surfaces, adapted vegetal diversity, minerality and urban furnitures strengthen the architecture of the building. Cutting through the lawn, low stone walls underline the horizontal line of the building and introduce a pace to the square. A walkway connects the park of Couzé to the neighborhood of la Picoterie and the townhall square becomes a pedestrian area.

设计说明

正如建筑纯净的线条一样，景观设计低调蓄地呈现出一片开阔的空间。各种植物和城市景观小品凸显并强化了建筑的风格。低矮的石墙穿过草坪，引至广场，强调建筑的水平线。一条走道将 Couzé 公园与附近的 la Picoterie 居民区相连，市政广场从而也变成了步行区。

Commercial Space
商业空间

Mixed Use
Open Space
Experience
Traffic Streamline

混合用途
开放空间
体验性
交通流线

COMMERCIAL AND PUBLIC LANDSCAPE 商业公共景观

CULTURAL SQUARE 文化广场　　URBAN SPACE 城市空间

Keywords 关键词
- Artwork 艺术品
- Signage 标识系统
- Functional 功能
- Specific Landscape 特色景观

Location: Northamptonshire, UK
Landscape Design: FoRM Associates

项目地点：英国北安普敦郡
景观设计：英国 FoRM 建筑事务所

Kettering Market Place
凯特林集市

Features 项目亮点

The design is not only people-oriented but also full of interest. It is a place for public events and meeting with friends, forming a pleasant and comfortable space.

该设计不但人性化，而且充满趣味，它既是公共活动的场地，又是聚会的场所，呈现出一个乐趣无穷的舒适空间。

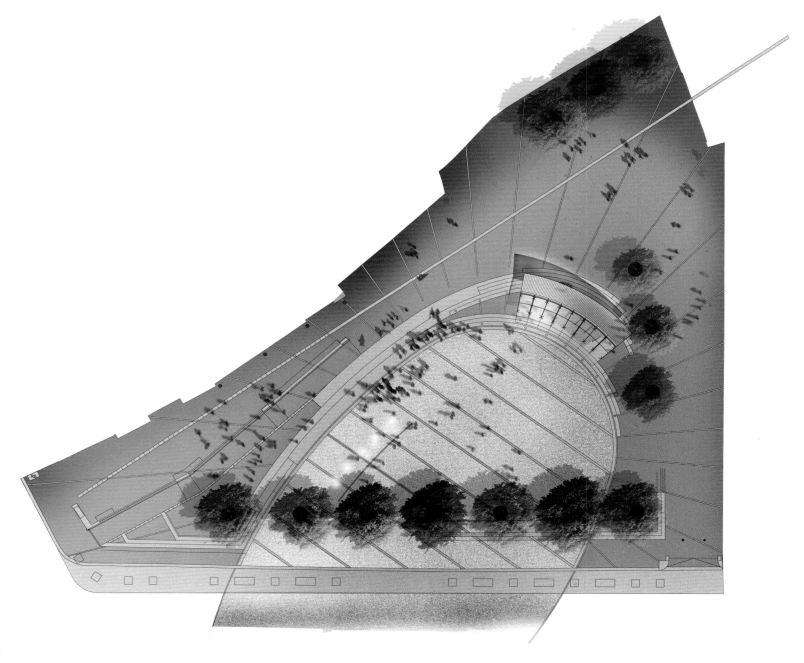

COMMERCIAL AND PUBLIC LANDSCAPE 商业公共景观 | CULTURAL SQUARE 文化广场 | URBAN SPACE 城市空间

Overview

The Market Place is one of the first initiatives to be completed with the aim of regenerating Kettering's town center and the surrounding area. It offers an attractive and fun place that is used to stage a range of public events throughout the year. It also provides an ideal setting to meet friends and to relax. One of the features of the Market Place is the fountains animated through lighting at night.

项目概况

该项目是为重振凯特林镇中心和周边地区所进行的第一批规划的内容之一，旨在设计一个可以举办大大小小公共活动的活力空间，同时也是跟朋友聚会放松的理想的环境。项目的特点之一是夜间照明灯下栩栩如生的人造喷泉。

| COMMERCIAL SPACE 商业空间 | OFFICE SPACE 办公空间 | HEALTHCARE AND EDUCATION 医疗教育 |

Design Description

The Market Place is revisioned as a vital public realm augmented by new landscape features and uses whilst retaining its original functions and character. The scheme gives priority to cyclists and pedestrians through subtle changes in surfacing. Overall access is improved through improved signage and accessibility measures. Increased seating opportunities, public artwork, water features and retail "spill-out" spaces bring life to the square.

设计说明

项目作为一个重要的公共领域空间，在保持其原有的功能和特点的基础上，增添了新的特色景观和用途。该方案优先考虑自行车流和行人在地面上的微妙变化。改进的标识系统和便捷的通行措施为旅客提供了更多的便利。座位数量、公共艺术品、水景和零售店面的增加，给广场增添了许多生机。

| COMMERCIAL AND PUBLIC LANDSCAPE 商业公共景观 | CULTURAL SQUARE 文化广场 | URBAN SPACE 城市空间 |

Keywords 关键词

Pragmatic Function 实用功能
Impressionistic Style 印象风格
Historical Building 历史建筑
Energy Conservation and Emission Reduction 节能减排

Location: New York, USA
Client: L & L Holding Company, LLC
Design: Landworks Studio, Inc., Boston

项目地点：美国纽约
客　　户：L & L Holding 有限责任公司
设　　计：波士顿 Landworks 工作室

200 5th Avenue New York City

纽约第五大街 200 号重建

Features 项目亮点

Energy efficiency, water reduction, daylighting, green building materials, and improved indoor air quality were a just few of the renovations planned to improve the building's environmental performance, not only to save on energy costs over time, but also to create better working conditions.

在原有建筑设施基础上，增加了有效能源消耗、水资源节能、采光节能以及绿色建筑的多方面设计，不仅仅在于节能减排，同时还旨在建立一个良好的适于企业发展的办公场所。

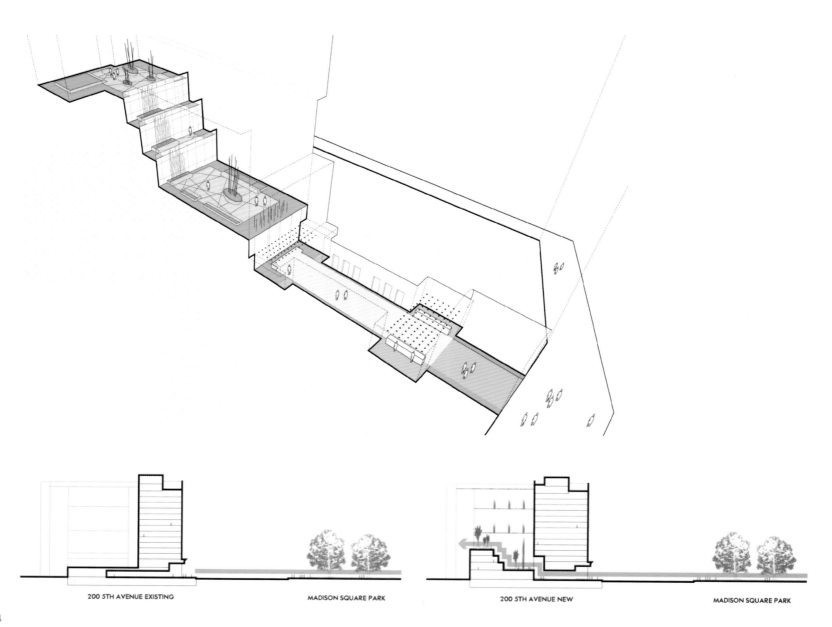

200 5TH AVENUE EXISTING   MADISON SQUARE PARK   200 5TH AVENUE NEW   MADISON SQUARE PARK

| COMMERCIAL SPACE 商业空间 | OFFICE SPACE 办公空间 | HEALTHCARE AND EDUCATION 医疗教育 |

Overview

The renovation of 200 5th Avenue, a historic landmark building adjacent to New York's Madison Square Park, breathed new life into its grand spaces and re-activated its civic presence. The contemporary courtyard is the centerpiece of this transformation, which contrasts old and new by leaving historic facades of the courtyard intact. The floating tray and lush plantings add brightness and texture to this landscape insertion and create a range of gathering areas and garden views.

项目概况

作为毗邻纽约麦德逊广场公园的一个历史性景观建筑，纽约第五大街200号的重建焕发出新的活力，增加了活动空间，同时也激发了市民的积极性。如今的庭院是整个改造项目的核心，保留了历史性的立面，融古今印象。浮盘与郁郁葱葱的植被是景观的亮点，增强了景观的纹理和层次，营造出一系列公共聚集区和花园景色。

COMMERCIAL AND PUBLIC LANDSCAPE 商业公共景观

CULTURAL SQUARE 文化广场

URBAN SPACE 城市空间

| COMMERCIAL SPACE 商业空间 | OFFICE SPACE 办公空间 | HEALTHCARE AND EDUCATION 医疗教育 |

COMMERCIAL AND PUBLIC LANDSCAPE 商业公共景观

CULTURAL SQUARE 文化广场

URBAN SPACE 城市空间

Design Description

Formerly known as the International Toy Center, 200 5th Avenue served for decades as a hub for toy manufacturers. One of many buildings that forged the architectural heritage of New York's Flatiron District, its U-shaped floor plate was considered highly innovative when it was built in 1909, due to the amount of natural light it brought into the building's interior and its creation of a courtyard for the building's residents.

In 2007, a new owner hired a team of designers to re-conceptualize and reinvigorate the fifteen-story commercial building by improving the quality of the workplace environment while preserving its historic architectural quality. Energy efficiency, water reduction, daylighting, green building materials, and improved indoor air quality were a just few of the renovations planned to improve the building's environmental performance, not only to save on energy costs over time, but also to create better working conditions. In addition to improving air quality by removing harmful toxins and circulating fresh air, the project also incorporated access to outdoor spaces such as the remodeled courtyard.

One of the main concepts the designer-developer team seized upon early in the design process was to visually connect the courtyard with Madison Square Park located at the intersection of Broadway and 5th Avenue. Employing a strategy of preservation augmented by a localized surgical intervention, the architect brought back the original 5th Avenue entry and replaced a solid interior wall with a fifteen-story glass curtain wall. These architectural renovations set the stage for visual connectivity to Madison Square Park throughout the lobby and extending into the interior courtyard.

设计说明

以世界玩具中心闻名的纽约第五大街200号近几十年来一直是众多玩具制造商的集聚之地。第五大道自1907年第五大街协会成立以来始终坚持高标准,使第五大道始终站在成功的顶峰。包揽众多货品齐全、受人喜爱的商店是第五大道的一个特色。

2007年,为了提升办公环境,阻止第五大街上的历史建筑的损耗,新的所有者聘请了一个专业的设计团队来重新设计和建设这栋15层的商业大楼。在原有的建筑设施基础上,增加了有效能源消耗、水资源节能、采光节能以及绿色建筑材质的多方面设计,不仅仅在于节能减排,同时还旨在建立一个良好的适于企业发展的办公场所。项目通过去除有害毒素,促进空气循环,合并入口与室外空间来提高空气质量。

其中一个重要的设计根据就是要在中央庭院与麦德逊广场之间建立有效的关联。通过有效的地表延伸,加强建筑的抽象派印象风格。通过对这个历史建筑的翻新,赋予新的现代风格特色,加强了建筑现代的工作实用性功能。

| COMMERCIAL SPACE 商业空间 | OFFICE SPACE 办公空间 | HEALTHCARE AND EDUCATION 医疗教育 |

COMMERCIAL AND PUBLIC LANDSCAPE 商业公共景观

CULTURAL SQUARE 文化广场

URBAN SPACE 城市空间

Keywords 关键词

Defined Space 有限空间
Wood Pergolas 木质藤架
Untouched Nature 原生态
Natural Experiences 自然体验

Location: Bodrum, Turkey
Architects: Tuncer Cakmakli Architects

项目地点：土耳其博德鲁姆
建筑设计：土耳其 Tuncer Cakmakli 建筑设计

Bodrum Primex Hotel

博德鲁姆 Primex 酒店

Features 项目亮点

The new design of interior spaces and the garden relies on natural experiences. Colors, smell, light, water and materials were composed as experiential elements.

花园的设计充满了绿色生机，通过对色、味、光、水、绿植等元素的精心打造，营造出自然、放松的舒适氛围。

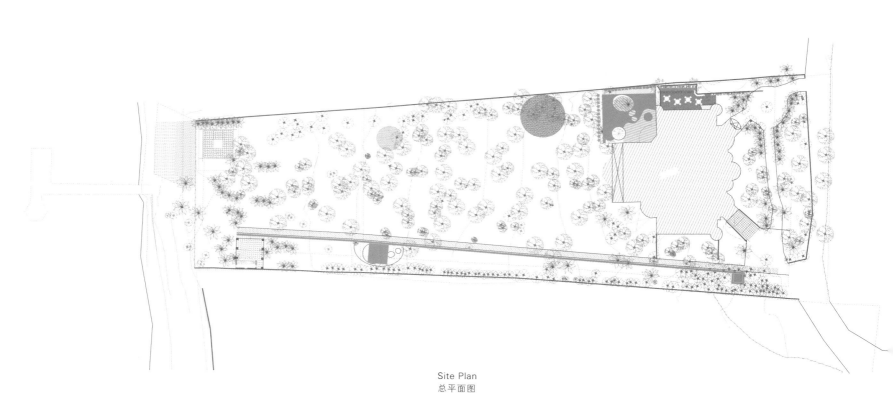

Site Plan
总平面图

| COMMERCIAL SPACE 商业空间 | OFFICE SPACE 办公空间 | HEALTHCARE AND EDUCATION 医疗教育 |

Overview

A boutique hotel on the coast of Aegean Sea, in untouched nature in Torba-Bodrum has been re-designed. A building that's been built but never used, was torn down to its foundations and re-designed.

The building is built on three levels, each 900m². The garden level features a large swimming pool, Turkish bath, sauna, game rooms, recreation rooms, a large industrial kitchen and servants rooms. In the entrance floor, the rooms are grouped around a large courtyard with a glass roof. There are the library, the dining room and living room. On the first floor there are two master and four VIP bedrooms. The entire roof is designed as a terrace.

项目概况

博德鲁姆 Primex 酒店是一座精品酒店，位于爱琴海岸，享有原生态的自然环境。在酒店未建之前，这里是一座从未使用过的建筑，建筑被拆除后，酒店取代了它的位置。

酒店共三层，每层900 m²。花园层包括大泳池、土耳其浴室、桑拿室、游戏房、娱乐室、厨房和酒店员工房间。入口层的所有房间围绕中庭排布，中庭上方是玻璃屋顶，这里有图书馆、餐厅和客厅。首层有两个大房间和四个VIP房。整个屋顶就是一个大露台。

COMMERCIAL AND PUBLIC LANDSCAPE 商业公共景观

CULTURAL SQUARE 文化广场

URBAN SPACE 城市空间

Design Description

The new design of interior spaces and the garden relies on natural experiences. Colors, smells, light, water and materials were composed as experiential elements. The various pergolas on the terrace, in the entrance and around the outside staircase are built from Akaschu wood, just like the shutters on the windows. They give protection against the sun and serve as a fragrant jasmine trellis. The building is thus gekrönnt and brought to the harmonious proportions.

The garden and the open spaces are clearly defined. A 150-meter-long road connecting the hotel entrance and the sea and an equally long water channel with a width of 43 cm along the way is used day and night for creating a fresh element on the way to the sea. Parallel to the road are orange, tangerine, and lemon trees planted, which fill the road with fragrance and color. On the other hand, the entire garden is covered with a carpet of grass, which invites you to relax under shady trees.

| COMMERCIAL SPACE 商业空间 | OFFICE SPACE 办公空间 | HEALTHCARE AND EDUCATION 医疗教育 |

设计说明

内部空间与花园的设计强调自然体验，设计师凭借多年经验，精选色、味、光、水等元素。露台、入口、室外楼梯等处的藤架由 Akaschu 木打造，充满着茉莉花的芬芳，如百叶窗一样，起到遮阳效果。

花园与开放空间界线分明。一条 150 m 的长道连接着酒店入口和爱琴海，旁边有条 43cm 宽、与该道等长的水渠，无论是在白天还是在夜里，涓涓流水都能给路人带来新鲜感。路旁还种植着橘子树和柠檬树，色香俱全。此外，整个花园绿草如茵、绿树成荫，是个休闲放松的好去处。

COMMERCIAL AND PUBLIC LANDSCAPE 商业公共景观

CULTURAL SQUARE 文化广场

URBAN SPACE 城市空间

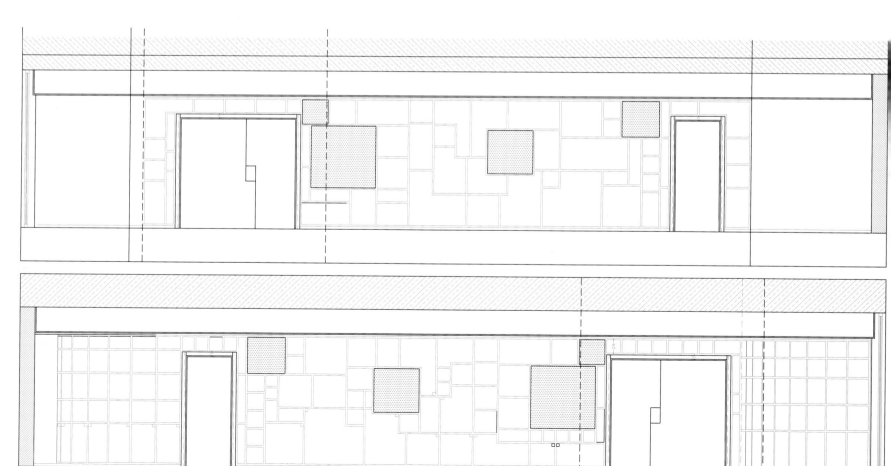

| COMMERCIAL SPACE 商业空间 | OFFICE SPACE 办公空间 | HEALTHCARE AND EDUCATION 医疗教育 |

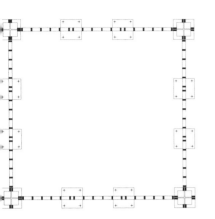
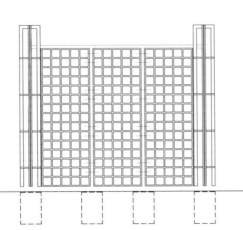
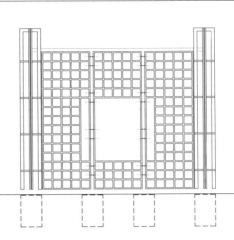
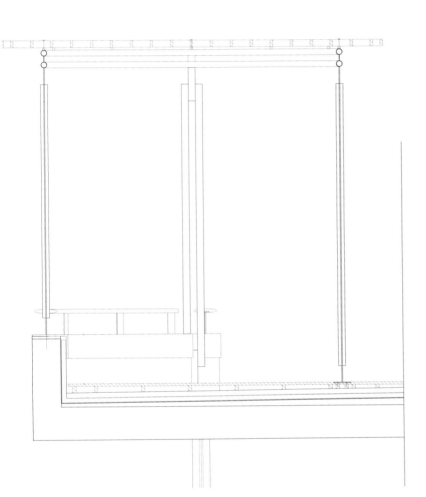

COMMERCIAL AND PUBLIC LANDSCAPE 商业公共景观

| CULTURAL SQUARE 文化广场 | URBAN SPACE 城市空间 |

| COMMERCIAL SPACE 商业空间 | OFFICE SPACE 办公空间 | HEALTHCARE AND EDUCATION 医疗教育 |

COMMERCIAL AND PUBLIC LANDSCAPE 商业公共景观

CULTURAL SQUARE 文化广场　　URBAN SPACE 城市空间

Keywords 关键词

Plants 植物
Lighting 灯光
Cooling Roof 冷却屋顶
Japanese Garden 日式花园

Location: Tokyo, Japan
Client: Nikken Sekkei, New Otani Hotel
Landscape Design: Keikan Sekkei Tokyo Co., Ltd.
Area: +/- 2,500 m²

项目地点：日本东京
客　　户：日建设计，新大谷酒店
景观设计：日本东京景观设计株式会社
面　　积：约2 500 m²

New Otani Hotel Roof Gardens
新大谷酒店屋顶花园

Features 项目亮点

Lighting is the significant design element of these two roof gardens, the architects built a peculiar lighting effect by traditional Japanese style which became the spectacular complement of this landscape design.

灯光是本案两个屋顶花园中最重要的设计元素，采用传统的日式风格，营造出别具一格的灯光效果，成为该景观设计的点睛之笔。

Overall Site Plan

| COMMERCIAL SPACE 商业空间 | OFFICE SPACE 办公空间 | HEALTHCARE AND EDUCATION 医疗教育 |

Overview

This project undertook the creation of roof gardens on the 17th floor roof (viewed from the restaurant) and the 3rd floor roof (top of the lower building), which were originally barren roofs, as part of the hotels renovations, necessary 40 years after the main building's construction. A unique spatial theme was developed defining the 17th floor roof as "Sky Garden" and the 3rd floor roof as "Relief Garden".

项目概况

项目包括酒店 17 层及 3 层的屋顶花园设计，酒店的屋顶花园之前毫无生机，40 年后客户对酒店进行翻新，并要求将 17 楼的屋顶花园定义为"空中花园"，将位于 3 楼的花园定义为"轻松花园"。

| COMMERCIAL AND PUBLIC LANDSCAPE 商业公共景观 | CULTURAL SQUARE 文化广场 | URBAN SPACE 城市空间 |

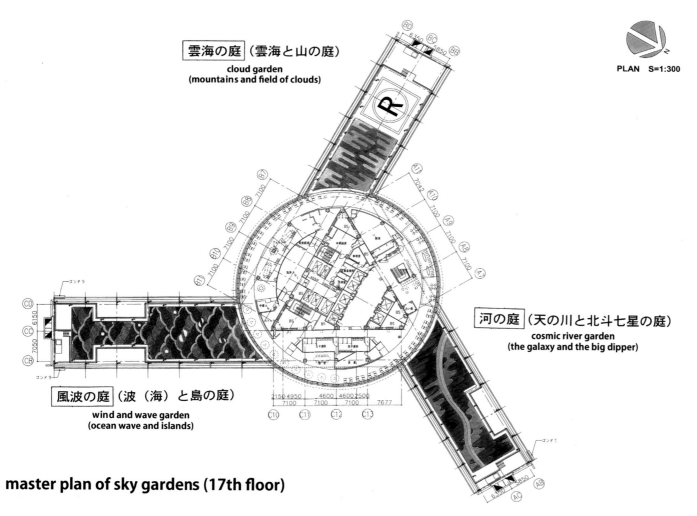

master plan of sky gardens (17th floor)

| COMMERCIAL SPACE 商业空间 | OFFICE SPACE 办公空间 | HEALTHCARE AND EDUCATION 医疗教育 |

COMMERCIAL AND PUBLIC LANDSCAPE 商业公共景观

CULTURAL SQUARE 文化广场 | URBAN SPACE 城市空间

Design Description

Due to the project's location issues such as weight restrictions, and countermeasures against wind blow were carefully considered in the context of the existing building structure to achieve a higher quality garden. Custom designed lighting is prominently featured throughout the gardens, becoming focal points among the plant material, much of which is ground cover plantings. These custom lighting fixtures have developed a great reputation due to their traditional Japanese styling-modest yet a significant component of the overall landscape.

The theme for the 3rd floor roof garden is "Relief Garden", a space for guests to feel peace of mind as they view Japanese nature from their guest rooms. The gardens communicate the Japanese perception of nature expressed through "Ka-Cho-Fu-Getsu" (Flower, Bird, Wind, Moon), with each modern Japanese garden representing a different theme, "Pine", "Bamboo", and "Plum", each of which carries symbolic beauty for Japanese.

These roof gardens are designed to help reduce the heat-island phenomenon in the Tokyo metropolitan area, and reduce stormwater runoff. As such the Tokyo Metropolitan Government contributed financial aid to the project due to its environmental considerations under the "Cool Roof Promotion Project". When combined with the existing Japanese garden of the New Otani Hotel and the surrounding green areas, these new roof gardens contribute both to making the hotel property more attractive as well as making the urban area of Tokyo more beautiful.

| COMMERCIAL SPACE 商业空间 | OFFICE SPACE 办公空间 | HEALTHCARE AND EDUCATION 医疗教育 |

设计说明

由于项目位置限制，设计师仔细考虑防风对策以打造更高质量的屋顶花园。灯光是该项目中最重要的设计元素，因此对植物的选择特别考究。由于采用了传统的日式风格，灯光效果别具一格——虽低调但在整个景观设计中乃点睛之笔。

3楼屋顶花园的设计主题是"轻松花园"，旨在为住客打造一个放松心情的空间，并且可以在客房内欣赏日式自然风景。花园设计通过花、鸟、风、月给人们传达对自然的感知，每一个日式花园代表一个不同的主题，松树、竹子和梅子都承载着象征日本人品格中的真善美。

屋顶花园的设计旨在降低城市热岛效应和雨水径流。该项目因其作为冷却屋顶的推广项目之一而得到东京政府拨款资助。结合新大谷酒店现存的日式花园与周边绿化空间，屋顶花园的设计将使酒店更具吸引力，同时使东京市容更美好。

COMMERCIAL AND PUBLIC LANDSCAPE 商业公共景观

CULTURAL SQUARE 文化广场

URBAN SPACE 城市空间

| COMMERCIAL SPACE 商业空间 | OFFICE SPACE 办公空间 | HEALTHCARE AND EDUCATION 医疗教育 |

| COMMERCIAL AND PUBLIC LANDSCAPE 商业公共景观 | CULTURAL SQUARE 文化广场 | URBAN SPACE 城市空间 |

Keywords 关键词

- Retail Park 零售公园
- Transplanting Plants 移栽植物
- Landscaping 造景
- Environment 环境

Location: Angers, France
Contractor / Client: Compagnie de Phalsbourg
Architect: Antonio Virga, AAVP
Landscape Design: Atelier Paul Arène
Total Area: 210,000 m²
Green Spaces Area: 62,000 m²

项目地点：法国昂热
客　　户：Compagnie de Phalsbourg
建 筑 师：Antonio Virga, AAVP
景观设计：法国 Atelier Paul Arène 工作室
总 面 积：210 000 m²
绿地面积：62 000 m²

France Atoll Retail Park

法国 Atoll 零售公园

Features 项目亮点

In order to step out of the conventional malls, extensive plants were imported to create an ambiance of forest and nature.

本设计将大量的植被引入场地，摒弃传统商业环境的景观风格，营造了一个透露着森林与自然气息的商业景观氛围。

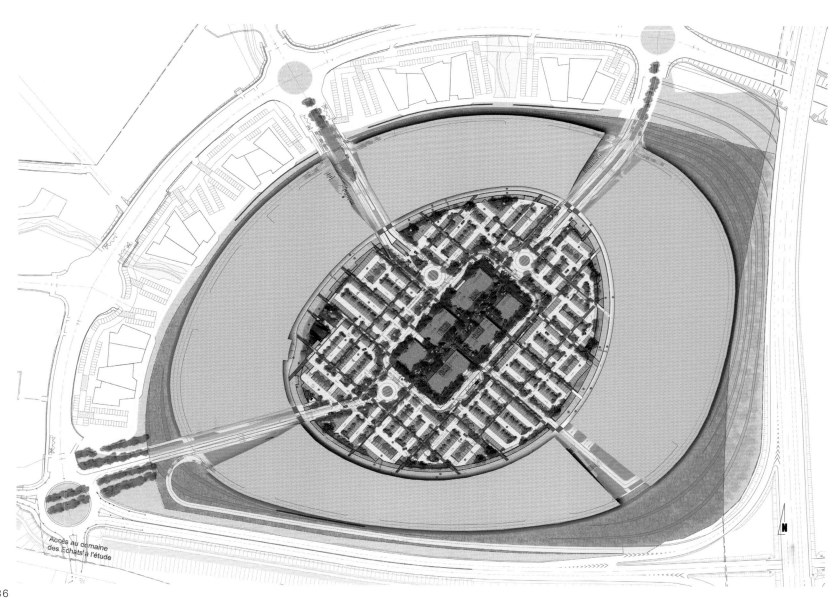

| COMMERCIAL SPACE 商业空间 | OFFICE SPACE 办公空间 | HEALTHCARE AND EDUCATION 医疗教育 |

Overview

Part of the economic development policy of the urban of Angers Loire Métropole, the Atoll is a retail park specialized in housing furnitures. The site is located in the municipality of Beaucouzé, on the edge of the urban and rural area. The importance given to the landscape design was the cornerstone of the project.

项目概况

作为响应昂热卢瓦尔河城市经济发展政策的产物，Atoll是一个专营房屋家具的零售公园。项目位于Beaucouzé的城乡结合处，因此，景观设计作为这个项目的奠基石至关重要。

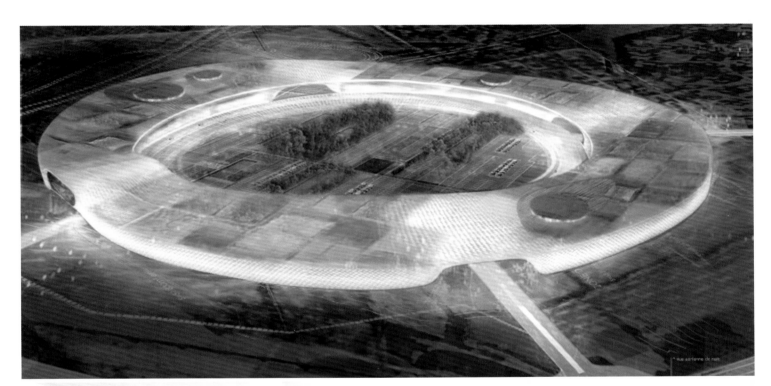

COMMERCIAL AND PUBLIC LANDSCAPE 商业公共景观

| CULTURAL SQUARE 文化广场 | URBAN SPACE 城市空间 |

Design Description

The concept is to step out of the conventional malls and their parking lots and to create an ambiance of forest and nature. In order to make this concept come true 6 acres of green spaces were created within this Ecoparc (15 acres with the outside of the ring) with 43,000 plants. Nearly 500 mature trees were imported and now contribute to the unique image of the project in terms of landscape design.

设计说明

设计想要摆脱传统的商场和停车场风格，打造出一种透露着森林与自然气息的氛围。为了实现这一想法，设计师在 Ecoparc15 英亩的外部圆形空间里空出6英亩的绿地，种植了近43 000 株植物，其中有500棵是移植过来的成熟树木，如今作为公园一景，给人的印象鲜明独特。

| COMMERCIAL SPACE 商业空间 | OFFICE SPACE 办公空间 | HEALTHCARE AND EDUCATION 医疗教育 |

| COMMERCIAL AND PUBLIC LANDSCAPE 商业公共景观 | CULTURAL SQUARE 文化广场 | URBAN SPACE 城市空间 |

| COMMERCIAL SPACE 商业空间 | OFFICE SPACE 办公空间 | HEALTHCARE AND EDUCATION 医疗教育 |

COMMERCIAL AND PUBLIC LANDSCAPE 商业公共景观

CULTURAL SQUARE 文化广场

URBAN SPACE 城市空间

| COMMERCIAL SPACE 商业空间 | OFFICE SPACE 办公空间 | HEALTHCARE AND EDUCATION 医疗教育 |

| COMMERCIAL AND PUBLIC LANDSCAPE 商业公共景观 | CULTURAL SQUARE 文化广场 | URBAN SPACE 城市空间 |

Keywords 关键词

Mixed-use 混合用途
Urban Space 城市空间
Waterscape Visual 水景视觉
Materials 材料

Location: Kansas City, Missouri, USA
Design: ZFG Architects LLP

项目地点：美国密苏里州堪萨斯城
设　　计：ZFG建筑师事务所

Crown Center Square
皇冠中心广场

Features 项目亮点

Base on the transformation and integration of the original simple urban space, the design creates a mixed-function space, highly tactile and visual effects.

本案基于对原有简单城市空间的改造和整合，将其设计成混合功能空间，具有极强的触觉和视觉效果。

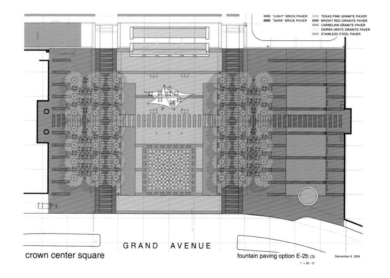

| COMMERCIAL SPACE 商业空间 | OFFICE SPACE 办公空间 | HEALTHCARE AND EDUCATION 医疗教育 |

Overview

Crown Center Square in Kansas City was transformed from a compromised mixed-use urban space with limited connectivity into an inviting urban plaza that promotes through-fare and respite time, supported by multiple water features and a tactile and visual high quality space.

The original square was conceived as a drop off vehicle court around a fountain at the center of the mixed-use development envisioned by Hallmark's founders in the 1960's. The complex evolved to include extensive office space, a shopping center, hotels, housing and public amenities.

项目概况

美国堪萨斯城的皇冠中心广场由一座联系性有限的折衷的混合用途城市空间改造成一个引人注目的城市广场，从而提高了票价和人们在此停留的时间，并布置了几处水景，形成触觉和视觉上都有很高质量的空间。

原先的广场只是一个围绕中心喷泉上下车的车站场，位于 Hallmark 创始人设想的混合用途城市空间的中心，修建于 20 世纪 60 年代。随后这里增加了办公空间、购物中心、饭店、住房和公共娱乐设施等。

COMMERCIAL AND PUBLIC LANDSCAPE 商业公共景观

CULTURAL SQUARE 文化广场

URBAN SPACE 城市空间

Design Description

The client brief was to bridge over Grand Avenue to engage retail activity, to restore the construction integrity of the space since it laid over a parking structure and to create a high quality public space to hold year-round themed events, all with the condition that the traditional City Christmas Tree Ceremony could not be disrupted by construction. In response, ZGF Architects, LLP developed a few design elements to enhance the existing context.

A central axial series of water features visually integrates the multiple office campus terraces to the retail across Grand Avenue. An iconic main water feature with programmable light, music and water shows is the central community focus. Flanking alleys of trees frame the central granite plaza and fountains, directly connect walkways between buildings and provide extensive tree canopy for shaded resting spaces. Paving materials integrate the order of the existing complex's structural patterns in subtle patterns and enrich the complex's palette with warmer colors in brick and granite.

Quality materials, strong construction measures and movable furnishings maximize flexibility of use and durability. The square is used year-round for public festivals so it must endure heavy vehicular traffic on a space over a parking structure, heavy pedestrian usage and year-round fountain usage.

| COMMERCIAL SPACE 商业空间 | OFFICE SPACE 办公空间 | HEALTHCARE AND EDUCATION 医疗教育 |

设计说明

改造的要求是在 Grand Avenue 大学上方架桥，连通各零售店，恢复这部分空间原先的整体性，并创造可举办全年主题活动的高质量公共空间。而且，施工活动不能打断该市传统的圣诞树庆典。为此，ZGF 事务所开发出一些新的设计要素来改善现状的脉络。

中轴线上一连串水景从视觉上整合了各个办公庭院和跨越 Grand Avenue 的零售店。通过程序化的灯光、音乐和水柱来表现的主水景区被改造为焦点。侧翼种植树木的小道界定了中央的花岗岩广场和喷泉，并直接连接了各建筑间的路径，其树木还为休息区提供了遮阴。铺路材料按照现有建筑的结构图案布置，十分精美，并以暖色调的砖石和花岗岩来丰富这一空间的色彩。

高质量的材料、有效的施工手段和可动性的装置扩展了空间的灵活性和耐久性。广场可举办整年的公共节日活动，所以必须耐受大量的车辆和步行者通行，其水景也是整年运行的。

COMMERCIAL AND PUBLIC LANDSCAPE 商业公共景观

CULTURAL SQUARE 文化广场

URBAN SPACE 城市空间

| COMMERCIAL SPACE 商业空间 | OFFICE SPACE 办公空间 | HEALTHCARE AND EDUCATION 医疗教育 |

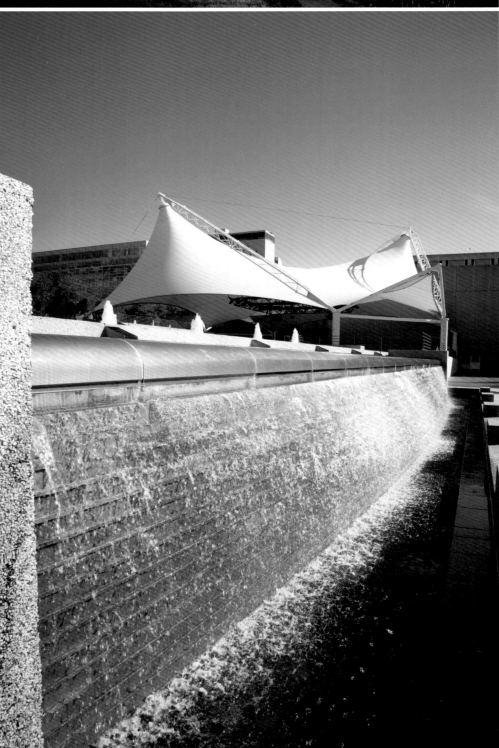

| COMMERCIAL AND PUBLIC LANDSCAPE 商业公共景观 | CULTURAL SQUARE 文化广场 | URBAN SPACE 城市空间 |

Keywords 关键词

Ramp 坡道地形
Landscape Terrace 景观阳台
Highveld Panorama 高原植被
Layout 布局

Location: Gauteng, South Africa
Client: Definite Properties
Landscape Design: GREENinc Landscape Architecture
Architects: Activate Architects

项目地点：南非豪登省
客　　户：Definite Properties
景观设计：南非 GREENinc 景观设计事务所
建筑设计：Activate 建筑事务所

Forum Homini Boutique Hotel

微型精品酒店景观

Features 项目亮点

It takes full advantage of the sloping terrain and mimics the entrance to a cave to integrate the public and private space together to create a definite hotel environment with characteristic balcony, vegetation and layout.

设计充分利用了坡地地形的优势，模仿洞穴入口的方式，将公共与私密空间进行有机的整合，塑造了一个阳台、植被及布局极富特色的酒店环境。

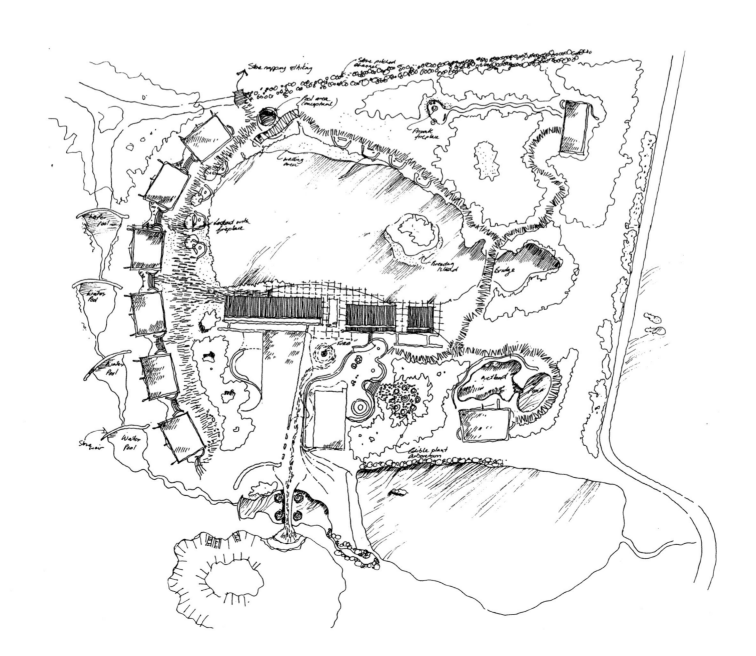

| COMMERCIAL SPACE 商业空间 | OFFICE SPACE 办公空间 | HEALTHCARE AND EDUCATION 医疗教育 |

Overview

The first buildings create a cut in the landscape, almost mimicking the entrance to a cave. Large hand-carved sandstone pillars march visitors down a ramp into the heart of the hotel complex. At the base of the ramp are the public buildings, reception, restaurant, and the reference library & conference facility are arranged around a lower courtyard and a higher landscape terrace.

项目概况

项目基地上的第一批建筑从景观创造入手，基本上模仿了洞穴入口的修建方式。大量手工雕刻的砂岩柱，把游客吸引到一个下坡道，由此便可进入酒店的心脏地带。公共建筑、接待处和餐厅散布在下坡道上，参考书阅览室及会议设施则布置在低矮的庭院和较高的景观阳台周围。

Design Description

The restaurant, lounge and library were positioned to have a view over an existing water-body. A pathway winds along the edge of this dam and connects the public buildings with the twelve suites, swimming pool and honeymoon suite. The rooms are partially submerged with veldgrass roofs and recede into the landscape. Once inside, floor to ceiling windows frame the highveld panorama, with a perennial stream in the foreground. As with the honeymoon suite, the presidential suite is set slightly apart from the other functions rooms of the hotel, and was placed to have the exclusive view of an existing wetland.

All disturbed areas were seeded with veldgrass, thereby creating a canvass for indigenous plants to be placed in between. The majority of the plant species selected is indigenous to the highveld and many of the species occurred naturally on the site. Exposed aggregate concrete pathways and stone clad retaining walls create a uniform link between the various buildings.

| COMMERCIAL SPACE 商业空间 | OFFICE SPACE 办公空间 | HEALTHCARE AND EDUCATION 医疗教育 |

设计说明

餐厅、酒吧和图书馆都以合适的位置分布，均可望到水景。沿着大坝边缘有一条蜿蜒的小路，连接了12间套房、游泳池和蜜月套房。房间屋顶部分有青草覆盖，形成了一道奇异的风景。进到里面，由顶及地的玻璃窗使人能够望到窗外的高地草原全景，和前面极有生气的常年流淌的小溪。和蜜月套房一样，总统套房与其他功能型酒店客房稍稍分开，专享湿地独特的景色。

所有相关的区域都种植了草坪，其中也引进了本土植物的种植。大部分挑选了高原本身土生土长的物种和当地自然生长的植物。裸露的混凝土小路和石砌挡土墙，将不同的建筑紧密联系在一起。

| COMMERCIAL AND PUBLIC LANDSCAPE 商业公共景观 | CULTURAL SQUARE 文化广场 | URBAN SPACE 城市空间 |

| COMMERCIAL SPACE 商业空间 | OFFICE SPACE 办公空间 | HEALTHCARE AND EDUCATION 医疗教育 |

COMMERCIAL AND PUBLIC LANDSCAPE 商业公共景观

CULTURAL SQUARE 文化广场

URBAN SPACE 城市空间

Keywords 关键词

Urban Space 城市空间

Garden Green 花园绿地

Sustainability 可持续性

Innovative 创新型

Location: Guangzhou, Guangdong, China
Landscape Design: SWA Group

项目地点：中国广东省广州市
景观设计：SWA 集团

Poly International Plaza

保利国际广场

Features 项目亮点

Firmly grasp the concept of sustainable development in the rapid urbanization, the design creates a modern aesthetic urban landscape environment through the integration of the surrounding garden green land and the building.

本案在快速的城市化进程中紧紧抓住了可持续发展的理念，通过建筑与周边一系列花园绿地的融合，营造了一个自然且透出现代美学感的城市景观环境。

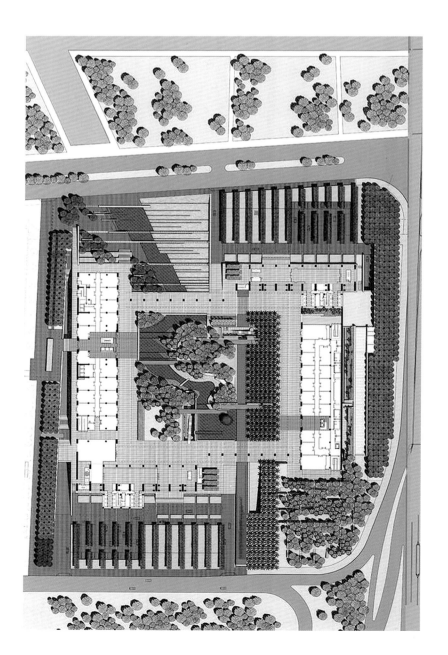

| COMMERCIAL SPACE 商业空间 | OFFICE SPACE 办公空间 | HEALTHCARE AND EDUCATION 医疗教育 |

Overview

Poly International Plaza is an innovative office and exhibition center development located in China's Guangzhou trade district. Sited along the Pearl River and adjacent to historic Pazhou Temple Park. The project presents a precedent toward integrating development with its site and context, embracing the place of garden and sustainability in the society's rapid move toward modernization. The result is a striking modern aesthetic that interacts efficiently and beautifully with the timeless elements of nature.

项目概况

坐落于珠江沿岸，毗邻历史悠久的琶洲塔公园，保利国际广场位于中国广州贸易区，是一个创新型办公和会展中心。项目根据其周边环境，在这个快速现代化的城市中首创了一个可持续的花园绿地，创造出一个大自然与现代美学相结合的经典案例。

| COMMERCIAL AND PUBLIC LANDSCAPE 商业公共景观 | CULTURAL SQUARE 文化广场 | URBAN SPACE 城市空间 |

| COMMERCIAL SPACE 商业空间 | OFFICE SPACE 办公空间 | HEALTHCARE AND EDUCATION 医疗教育 |

| COMMERCIAL AND PUBLIC LANDSCAPE 商业公共景观 | CULTURAL SQUARE 文化广场 | URBAN SPACE 城市空间 |

| COMMERCIAL SPACE 商业空间 | OFFICE SPACE 办公空间 | HEALTHCARE AND EDUCATION 医疗教育 |

COMMERCIAL AND PUBLIC LANDSCAPE 商业公共景观

CULTURAL SQUARE 文化广场

URBAN SPACE 城市空间

Keywords 关键词

Headquarters 总部属性
Industrial Environment 工业环境
Traffic Line 交通流线
Landscape Space 景观空间

Location: Saint Berthevin, France
Contractor/Client: Gruau
Project Management: Atelier Paul Arène
Total Area: 76,000 m²

项目地点：法国圣贝特万市
客　　户：Gruau
项目管理：法国 Paul Arène 工作室
总 面 积：76 000 m²

Gruau Industrial Park
Gruau 工业园

Features 项目亮点

The main characteristics of the site are its numerous buildings and an important need for parking space. A new traffic plan after reorganizing and the superior landscape form another characteristic.

大量建筑和重要的停车位需求是该地块的特色，项目改造后的交通规划与高品质的景观成为了园区空间设计的新亮点。

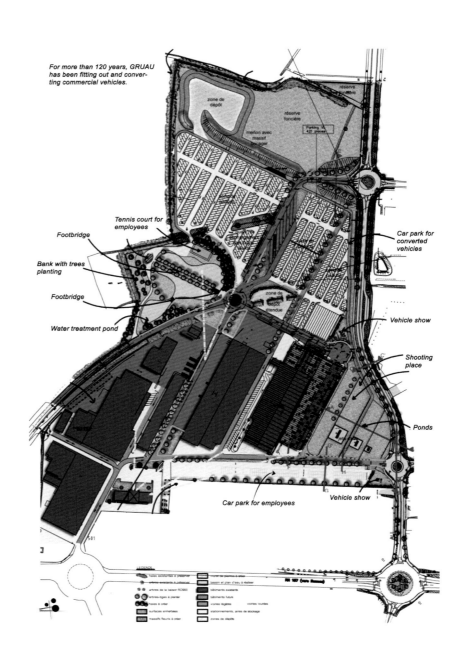

| COMMERCIAL SPACE 商业空间 | OFFICE SPACE 办公空间 | HEALTHCARE AND EDUCATION 医疗教育 |

Overview

The Gruau headquarters are located in the municipality of Saint Berthevin, France. From an original surface area of 19 acres (about 76,000m^2) to extended area of 34 acres (about 137,000m^2). The main characteristics of the site are its numerous buildings and an important need for parking space. This particular context required a specific approach that took place with a new traffic plan, the set-up of a commercial front line along with the development of a space dedicated to receptions and creativity. The main objective of the project was to create a high quality landscape development for the Gruau headquarters.

项目概况

Gruau 总部位于法国圣贝特万市,原有面积 19 英亩(约 76 000 m^2),现扩大至 34 英亩(约 137 000 m^2)。大量建筑和重要的停车位需求是该地块的特色。基于地块现有的特殊情况,设计师为它打造了一种新的交通规划,也营造出一个商业空间专门用于接待和创造力的培养。不过该项目的主要目标是为 Gruau 总部创造一个高品质的景观空间。

| COMMERCIAL AND PUBLIC LANDSCAPE 商业公共景观 | CULTURAL SQUARE 文化广场 | URBAN SPACE 城市空间 |

| COMMERCIAL SPACE 商业空间 | OFFICE SPACE 办公空间 | HEALTHCARE AND EDUCATION 医疗教育 |

COMMERCIAL AND PUBLIC LANDSCAPE 商业公共景观

CULTURAL SQUARE 文化广场　　URBAN SPACE 城市空间

Keywords 关键词
- Plant 植栽
- Low-cost 低成本
- Environment 环境
- Spanish-style 西班牙风格

Location: Madrid, Spain
Client: Parque Empresarial Cristalia, S.L.
Design: RTKL Associates Inc.
VP-in-Charge: Carlos Martinez
Area: 90,000m²
Photographer: David Whitcomb

项目地点：西班牙马德里
客　　户：Parque Empresanal Cristalia S.L.
建筑／景观设计：RTKL 建筑事务所
项目负责人：卡洛斯·马丁内斯
面　　积：90 000 m²
摄　　影：大卫·惠特科姆

Cristalia Business Park
Cristalia 商业园

Features 项目亮点

Inspired from the Spanish landscape and the local environment, the design creates a rich landscape, environment-friendly and Spanish-style Business Park.

设计从西班牙景观和当地环境获得灵感，建造出景观丰富、环境友好的西班牙风情商业园区。

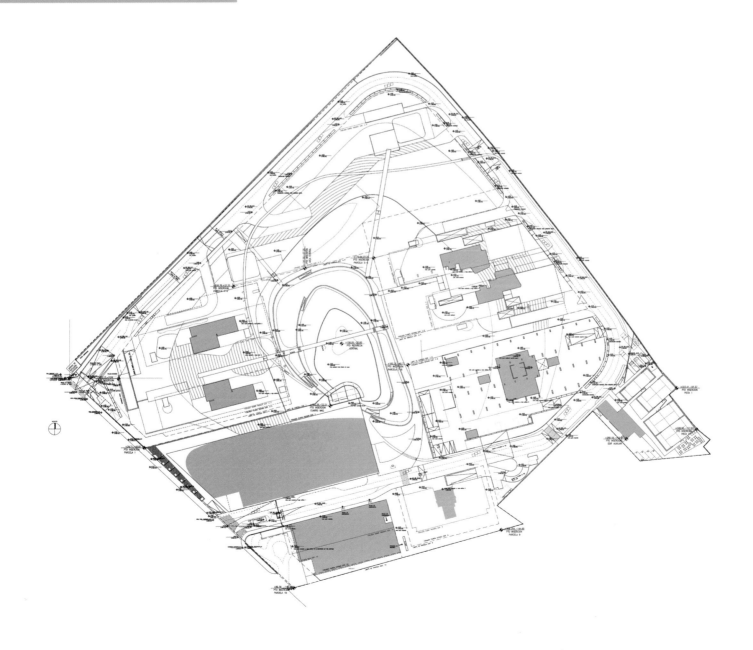

| COMMERCIAL SPACE 商业空间 | OFFICE SPACE 办公空间 | HEALTHCARE AND EDUCATION 医疗教育 |

Overview

The 90,000m² Cristalia Business Park, located in Spain's capital city, consists of eleven buildings, including office towers, restaurants and a hotel.

项目概况

Cristalia 商业园位于西班牙首都,占地面积 90 000 m²,共 11 栋建筑,包括几座办公大楼、几家餐厅和一个酒店。

COMMERCIAL AND PUBLIC LANDSCAPE 商业公共景观

CULTURAL SQUARE 文化广场

URBAN SPACE 城市空间

Design Description

The landscape design scheme emphasises interior space and uses landscape and garden green area, restaurants and park terraces to create a quiet and attractive environment for pedestrians and workers inside the buildings.

Moreover, RTKL established a series of space that are inspired by the Spanish landscape and vernacular built environment: a central water feature with an adjacent grass surface reminds one of the river shores, the rhythmic placement of olive trees reminds one of Southern Spain, and the geometric rows of seasonal scent plants and flowering shrubs remind one of the cultivated fields of Spanish countryside. RTKL selected local species of plants and trees to reduce cost maintenance, and environmental impact.

| COMMERCIAL SPACE 商业空间 | OFFICE SPACE 办公空间 | HEALTHCARE AND EDUCATION 医疗教育 |

设计说明

观设计方案强调内部空间，运用园林花园绿地、餐厅和公园露台，为行人和工作人员创造一个安静、舒心的环境。

外，设计人员根据从西班牙景观和乡土建筑环境获得的灵感设计了多处小景点。如一个让人如同身处河畔的带草地的中央喷泉、一处有西班牙南部风情的橄榄树、一条几何线形伴有季节香味的植物和开花灌木的西班牙乡村小道。设计人员采用当地的植物和树木，以降低建造和维护成本，减少对环境的影响。

COMMERCIAL AND PUBLIC LANDSCAPE 商业公共景观

CULTURAL SQUARE 文化广场

URBAN SPACE 城市空间

| COMMERCIAL SPACE 商业空间 | OFFICE SPACE 办公空间 | HEALTHCARE AND EDUCATION 医疗教育 |

COMMERCIAL AND PUBLIC LANDSCAPE 商业公共景观

| CULTURAL SQUARE 文化广场 | URBAN SPACE 城市空间 |

| COMMERCIAL SPACE 商业空间 | OFFICE SPACE 办公空间 | HEALTHCARE AND EDUCATION 医疗教育 |

123

COMMERCIAL AND PUBLIC LANDSCAPE 商业公共景观

CULTURAL SQUARE 文化广场 | URBAN SPACE 城市空间

Keywords 关键词

Architectural Facade 表皮肌理
Pavement Material 铺地材料
Platform Space 平台空间
Indigenous Vegetation 特色植物

Location: 717 Bourke Street, Docklands, Melbourne, Victoria, Australia
Client: ProBuild & PDS Group
Team: ASPECT Studios Landscape Architecture (lead consultant)
　　　Metier 3 Architecture
　　　Aurecon Structural, Civil, Hydraulic, Electrical Engineering
　　　Blythe Sanderson Access and Mobility
　　　Probuild Constructions Builder
Area size: 2,185m²
Photographer: Andrew Lloyd

项目地点：澳大利亚维多利亚州墨尔本海港区 717 Bourke 商业街
业　　主：ProBuild & PDS Group
设计团队：ASPECT Studios 澳派景观设计工作室
　　　　　Metier 3 Architects 建筑设计公司
　　　　　Aurecon 结构、水电工程设计公司
　　　　　Blythe Sanderson 无障碍设计公司
　　　　　Probuild 施工单位
项目面积：2 185 m²
摄 影 师：Andrew Lloyd

717 Bourke Street
717 Bourke 商业街

Features 项目亮点

The integration of strong landscape forms and sophisticated urban design has created a connective, multifunctional and successful urban space.

独特的景观形式与先进的城市设计相结合，打造出一处交通流畅、功能多样的成功的都市空间。

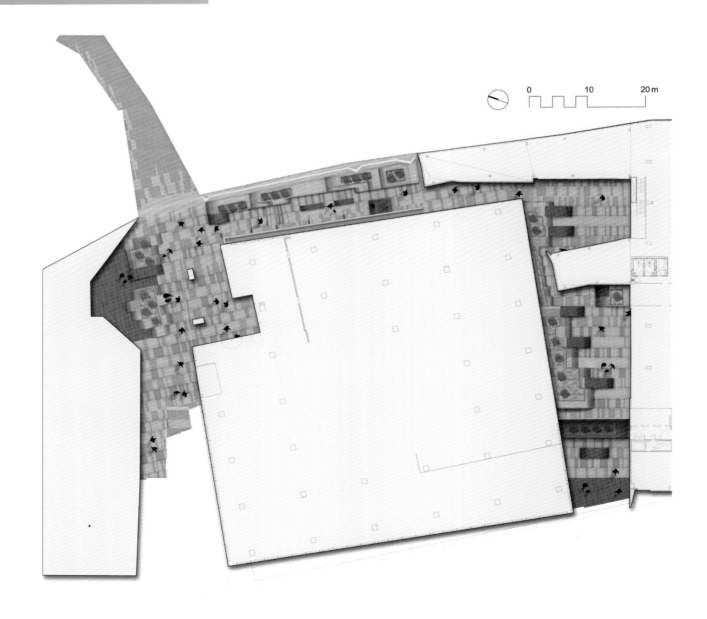

| COMMERCIAL SPACE 商业空间 | OFFICE SPACE 办公空间 | HEALTHCARE AND EDUCATION 医疗教育 |

COMMERCIAL AND PUBLIC LANDSCAPE 商业公共景观

CULTURAL SQUARE 文化广场

URBAN SPACE 城市空间

Overview

The streetscape, pocket park, podium forecourt and the courtyards of a mixed use development in Melbourne's Docklands, create a multileveled urban landscape realm that both continues ASPECT Studio's exploration of the urban design possibilities of inner city built form, and extends in new ways the idea of the corporate plaza. The design seeks to create dialogue between architecture, the layering of horizontal flows the site requires, and the significance of site, and to form a place of warmth, niches, and unpredictable spaces.

Docklands is the newest urban extension of the city of Melbourne, the transformation of industrial port land into a new urban precinct, a redevelopment of a type similar to Canary Wharf in London and Hafen Hamburg in Germany. The site is at the active edge between the new development and the historic city grid, immediately adjacent to the major bridge link from the city's largest train station—Southern Cross Station in the CBD to Etihad Stadium.

项目概况

墨尔本海港区新的商业综合体项目的设计对象是包括商业街、小公园、平台广场以及庭院花园在内的多层次都市空间。此项目是澳派景观设计工作室对于如何打造更成功的城市空间的思考与探索，包括城市空间的构造形式以及新的写字楼广场的功能等。设计使景观与建筑形成对话，并巧妙地利用现状地形营造出一系列的平台空间，形成温馨舒适、尺度宜人，甚至让人意想不到的空间形式。

海港区是墨尔本最新打造的城市扩展区域，将旧的工业港口转变为新的城区，类似于伦敦的金丝鸟港以及德国汉堡港。项目位于墨尔本新城区与历史老城的交汇地带，紧邻该市最大的火车站——CBD南极星站和Etihad体育场之间。

| COMMERCIAL SPACE 商业空间 | OFFICE SPACE 办公空间 | HEALTHCARE AND EDUCATION 医疗教育 |

Design Description

On a broad scale, the design's new footbridge forms a link to a series of stylised, intimate podium landscapes that cascade to street level. As the first new urban connection from the CBD grid to the street network of Docklands, these landscapes facilitate improved pedestrian traffic and increased activation between Melbourne's CBD and the predominantly isolated Docklands precinct.

The podium landscapes are seamlessly integrated with the imposing architectural façade and enveloping canopy form, through the creation of a topographical "carpet" that spreads outwards from the building, blurring the interface between public and private, active and passive building and landscape. The design deconstructs the typical 90 degree relationship between landscape and building by creating a language of tectonic landscape forms that operate as ramps, seats, garden beds, platforms and decks. The paving wraps up into the lines of the building itself.

In an acknowledgement of the site's location at the edge of former swamp and adjacent to the historic and now razed Batman's Hill, the topographic "carpet" is both shifted upwards to form occupiable plinths and platforms, and drawn apart to reveal the hidden indigenous vegetation within. An orchestrated series of stairs rising 14 metres from street level, and enveloped by the grand canopy form, establish private podiums, office courtyards, and commercial spaces, generate pedestrian engagement, and further assert the historic presence of Batman's Hill. The multiple stories operating in this design show a rich design poetics and the multiple nature of space.

Timber platforms conceal irrigated podium planter boxes connected into the stormwater recycling system, and which contain copses of drought tolerant Banksia and small Eucalypts with under plantings of native grasses and strappy shrubs. The platforms provide physical separation between the primary

COMMERCIAL AND PUBLIC LANDSCAPE 商业公共景观

CULTURAL SQUARE 文化广场 | URBAN SPACE 城市空间

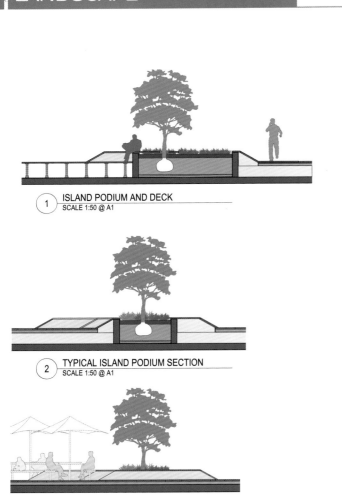

1. ISLAND PODIUM AND DECK
SCALE 1:50 @ A1

2. TYPICAL ISLAND PODIUM SECTION
SCALE 1:50 @ A1

3. TYPICAL ISLAND PODIUM ELEVATION
SCALE 1:50 @ A1

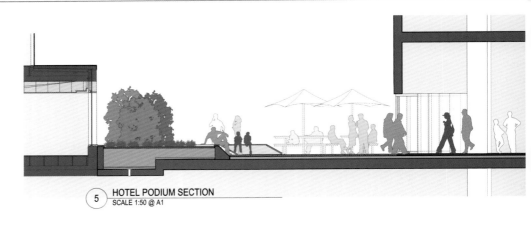

5. HOTEL PODIUM SECTION
SCALE 1:50 @ A1

6. POCKET PARK SECTION
SCALE 1:50 @ A1

| COMMERCIAL SPACE 商业空间 | OFFICE SPACE 办公空间 | HEALTHCARE AND EDUCATION 医疗教育 |

circulation link and an outdoor café zone, with the timber deck wrapping up the walls to form the canopy that flows down to street level. Constructed of pre-ordered Australian hardwood timber and stockpiled for several months to leach out tannins, these timber constructions are detailed to eliminate visible fixing. Their warm natural tones balance the cooler tones of the hardscape "carpet", create a sense of calm, and draw the visitor to this place of repose.

The "carpet" consists of bands of paving-granite, bluestone, and two colour variations of splitstone-broken down into a geometric mosaic pattern that enhances the articulation of the façade and creates a shifting, finely scaled and textured surface. The use of bluestone extends the city of Melbourne's palette and character, and brings to the design language of the functionality, identity and history fusion.

Additional white stone surfacing to all vertical surfaces for safety and legibility is continued into a faceted "white snake" wind amelioration wall that shields the main stepped podium landscape. Extensive wind modelling informed the design of this wall element which emerges from a series of outdoor "sunroom" terraces.

The integration of strong landscape forms and sophisticated urban design has created a connective, multifunctional and successful urban space, a corporate plaza for the public, and a place of significance within the new Docklands precinct.

| COMMERCIAL AND PUBLIC LANDSCAPE 商业公共景观 | CULTURAL SQUARE 文化广场 | URBAN SPACE 城市空间 |

设计说明

从大范围来说，新建的人行桥和街道地面通过一系列充满设计感的、层层叠叠如瀑布一般的小平台连接在一起。作为联系墨尔本 CBD 旧城区与海港区的街道项目之间的第一条纽带，这一座人行桥大大改善了人行交通，并加强了旧城区与位置相对偏僻的海港区的联系。

设计师将表现感很强的建筑外表面肌理运用到了户外的铺地上，如同一张地毯，将建筑与景观无缝地连接在一起，亦将突出的平台包裹在里面，融合公共与私有空间，动态与静态空间。设计打破了景观与建筑常规的 90 度直角的空间关系，创造出一种独特的景观语言，成功地打造出一系列商业景观功能空间，包括梯形的景观地形、座椅、种植池和平台等。铺装与建筑本身的线条融为一体。

场地旁原本是一片沼泽，还紧邻一座现在已被夷为平地的巴曼斯山，为了回应历史，整个景观地形被设计成"地毯"一样向上方延伸，形成一个个有趣的平台空间，并配置本地的特色植物。设计师精心设计了一组 14 m 高的平台，并用巨大的树冠形结构笼罩起来，以强调巴曼斯山存在过的历史真实性。具体来说，这些平台空间有私人休息平台、写字楼庭院花园、商业空间、公共人行空间等。整个多层的景观设计体现出空间的韵律感和此处丰富的历史地貌。

种植池巧妙地隐藏在景观平台之后，种植池下方设有起到过滤作用的碎石带，多余的浇灌用水可以在过滤后流入整个项目的雨水系统。选用的景观植物有非常耐旱的当地植物如澳洲山龙眼与柠檬桉等，下层是本地的灌木与地被植物。这些平台空间巧妙地区分出公共人行区域与室外阳光咖啡座的空间；木平台的木板翻卷越过墙体，变化为下方街道的木廊架。所用的木条是澳大利亚本土的硬木，在施工之前需要进行特别处理除去丹宁酸，以保证木条的耐久性。设计师仔细推敲结构细节，保证木条连接件都被隐藏起来，保证更好的视觉效果。温暖的木质材料与相对硬朗的铺地相结合，营造出更为宁静、舒适的空间感受，吸引更多的市民来到此处。

南北向的景观铺地所用的材料是花岗岩条石、青石与两种不同色调的石材，创造出闪烁精细的材质表面，形成精美的马赛克拼花图案，呼应建筑立面的表情。设计师沿用了墨尔本旧城区主要使用的材料——青石，创造出功能合理、独具个性，并充满历史风情的设计语汇。

场地内所有的立面都用白色小石子饰面，既保证安全性也塑造出独特的风格，这样的饰面也延续到一面保护整个平台空间的三维挡风墙上。这面挡风墙是经过全面的风力测试而最终确定的，同时保证各个平台空间拥有良好的采光特性。

独特的景观形式与先进的城市设计相结合，打造出一处交通流畅、功能多样的成功的都市空间，既是公共的商业广场也是海港区具有重要意义的城市核心。

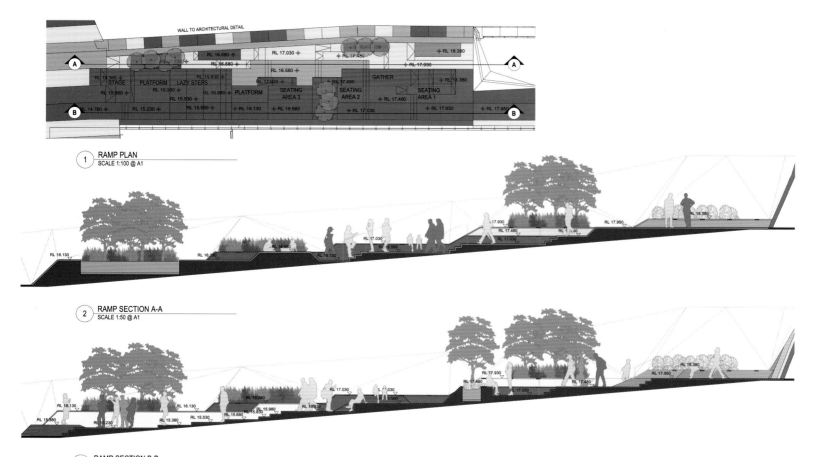

| COMMERCIAL SPACE 商业空间 | OFFICE SPACE 办公空间 | HEALTHCARE AND EDUCATION 医疗教育 |

COMMERCIAL AND PUBLIC LANDSCAPE 商业公共景观
CULTURAL SQUARE 文化广场 | URBAN SPACE 城市空间

Keywords 关键词
- Site Atmosphere 场所氛围
- Space Design 空间设计
- Plant Landscape 植物造景
- Material 材料

Location: Amsterdam, Holland
Design Team: LOUP & Co Outdoor Luxury
Surface: 350 m²

项目地点：荷兰阿姆斯特丹
设计团队：法国 LOUP & Co Outdoor Luxury
面积：350 m²

The "Grand" Hotel Amsterdam
阿姆斯特丹索菲特大酒店庭院

Features 项目亮点

Designers reorganize and improve the external and internal of this courtyard with a variety of plants, creating a contemporary and comfortable living space.

该庭院的设计通过对内外空间的重组和改进，打造出绿植环绕，舒适惬意的当代生活空间。

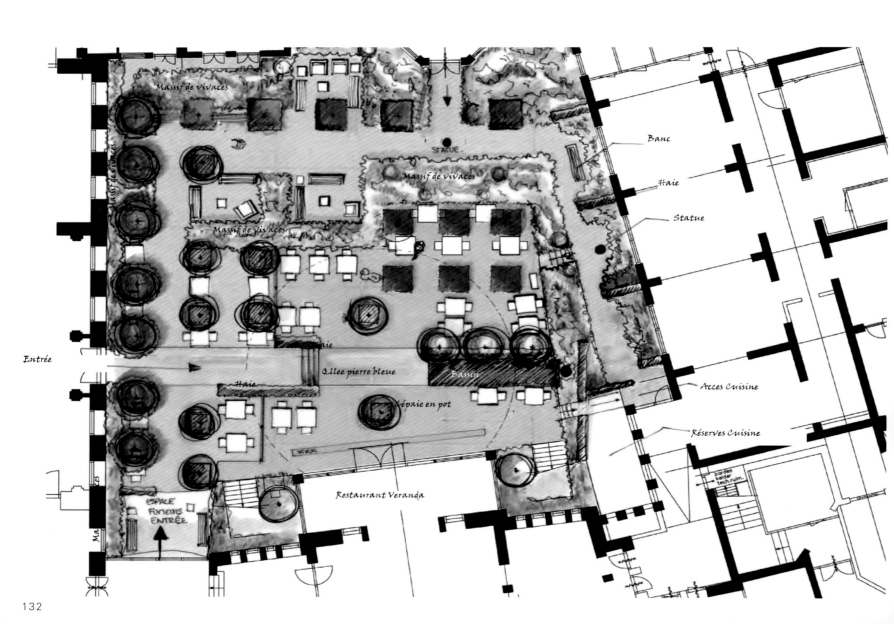

| COMMERCIAL SPACE 商业空间 | OFFICE SPACE 办公空间 | HEALTHCARE AND EDUCATION 医疗教育 |

COMMERCIAL AND PUBLIC LANDSCAPE 商业公共景观 | CULTURAL SQUARE 文化广场 | URBAN SPACE 城市空间

TERRASSE

| COMMERCIAL SPACE 商业空间 | OFFICE SPACE 办公空间 | HEALTHCARE AND EDUCATION 医疗教育 |

Overview

There are some problems to be solved in this grand hotel: polluted on site soil, difficult access, secluded garden and inconsistent living spaces.

项目概况

该项目为阿姆斯特丹索菲特大酒店的庭院设计，项目存在的问题包括土壤污染，入口难近，花园隐蔽，居住空间不一致。

COMMERCIAL AND PUBLIC LANDSCAPE 商业公共景观

CULTURAL SQUARE 文化广场

URBAN SPACE 城市空间

| COMMERCIAL SPACE 商业空间 | OFFICE SPACE 办公空间 | HEALTHCARE AND EDUCATION 医疗教育 |

Design Description

Objectives: Creating contemporary living units, restaurant's dinning space, and lounges; restructing of living spaces and circulation planning; creating of vegetal lodges used as workspace; recycling existing floor covers; modernizing the space as per Sybille De Margerie's interior refurbishment and decoration; playing with levels; refreshing and enhancing exterior spaces' facade; designing spaces lighting automating watering; ensuring all year long vegetal beauty; equipping outdoor sound system.

Materials : Recycled bricks on field, blue stone tiling and epoxy painted steel mirror basin, tailor made fibrocement trays.

Vegetal elements: Buxus sempervirens bowls and cushions, osmanthus clumps, x burkwoodii escalade, carpinus betulus preformed cube rods, grasses, anemones, hydrangeas.

设计说明

设计目标：打造当代生活空间、餐厅就餐空间与休息室；重组生活空间与交通路线；打造绿植环绕的门房，用作工作空间；回收现有地板层；按照Sybille De Margerie室内翻新与设计的标准改进该空间；保证层次分明；更新外部空间；改进建筑外墙；设计空间照明；设置自动浇水系统；确保植物美颜闪亮一整年；配备室外音响系统。

材料：再生砖、蓝色石材瓷砖、环氧涂钢盆镜、特制石棉水泥托盘。

植物：锦熟黄杨、木犀属植物、鹅耳枥、草、海葵、绣球花等。

| COMMERCIAL AND PUBLIC LANDSCAPE 商业公共景观 | CULTURAL SQUARE 文化广场 | URBAN SPACE 城市空间 |

RESTAURANT

| COMMERCIAL SPACE 商业空间 | OFFICE SPACE 办公空间 | HEALTHCARE AND EDUCATION 医疗教育 |

Office Space
办公空间

Open Space
Flexible Layout
People Orientation
Energy Efficiency

开敞空间
灵活布局
人性化
节能减排

| COMMERCIAL AND PUBLIC LANDSCAPE 商业公共景观 | CULTURAL SQUARE 文化广场 | URBAN SPACE 城市空间 |

Keywords 关键词

Plantation 植栽
Environment 环境
Sustainability 可持续性
Desert Landscape 沙漠景观

Location: Rancho Mirage, California, USA
Client: The Annenberg Foundation Trust at Sunnylands
Landscape Design: The Office of James Burnett

项目地点：美国加利福尼亚州兰乔米拉奇
客　　户：阳光岛安信托基金会
景观设计：James Burnett 设计公司

Sunnylands Center & Gardens in Rancho Mirage, CA

加州兰乔米拉奇阳光岛花园中心

Features 项目亮点

Designers arranged variant desert plants in painterly way to create a new ecological aesthetic landscape.

设计师将种类繁多的沙漠植物按照绘画的方式栽植，打造出一个生态和美学意义并举的独特景观。

| COMMERCIAL SPACE 商业空间 | OFFICE SPACE 办公空间 | HEALTHCARE AND EDUCATION 医疗教育 |

Overview

The landscape designer created a living landscape that respects the character of the Sonoran Desert and demonstrates a new ecological aesthetic landscape in the arid southwest.

项目概况

景观设计师设计出一个富有活力的景观氛围，不仅反映出所诺兰沙漠的地质特征，同时也成为干旱的西南部一个富有生态和美学意义的独特景观。

COMMERCIAL AND PUBLIC LANDSCAPE 商业公共景观

CULTURAL SQUARE 文化广场

URBAN SPACE 城市空间

Design Description

Because of its location in the desert, sustainability figured prominently into discussions about the nature of the project. The landscape designer sculpted the earth and used plants in a painterly fashion across the 15 acre site. Trees were carefully positioned throughout the site to ensure that ample shade was provided and great care was given to the visual composition of understory plantings.

In collaboration with the design, the building was carefully sited to frame panoramic views of the mountains beyond from the Center's main entry and lobby. A continuous terrace across the west side of the building extends the Center's café to the landscape and accommodates special events. Two stainless steel fountains within the terraces complement the crisp architectural composition, mirror the expansive desert sky, lower the ambient temperature and create the soothing sound of moving water.

Sized specifically to support special events required for programming, a circular lawn is the central organizing feature of the rear garden. Framed by a double row of "Desert Museum" Palo Verdes, its perimeter walk connects guests to a series of private gardens that feature quite seating nooks, rich desert plantings and a labyrinth for contemplation. Paths from the rear garden lead visitors through a rich and varied botanical collection of desert plants that passes through the front garden, along the perimeter of the restored habitat and back to the Center.

| COMMERCIAL SPACE 商业空间 | OFFICE SPACE 办公空间 | HEALTHCARE AND EDUCATION 医疗教育 |

设计说明

鉴于项目地处沙漠的特殊性，设计师在考虑项目的未来发展方面着重强调了可持续性。景观设计师将地表进行重新雕琢，在 15 英亩（约 60 703 m²）的土地上将各种植物按照绘画的方式进行编排，使树木的栽植在不影响低矮植物生长的前提下提供充足的阴凉。

建筑设计师与景观设计师联手使中心的主入口和大厅处可以尽享远山的全景。西面连续的平台将中心餐厅延伸到景观之中，可以用来举办特别的盛宴。平台中央两座不锈钢喷泉形成建筑的有机补充，同时还起到反射天空景色，降低环境温度以及创造舒缓水声的作用。

中心花园的建立是为了承办与基金会有关的特殊仪式，圆形草坪成为后花园内的主要特色。草坪两旁各有一座沙漠博物馆，沿环草坪的边缘前进可以引领宾客到达一系列私人庭院，内设角落座椅、种类繁多的沙漠植物以及用于沉思的迷宫等。参观者沿着后花园内的小径前行可以欣赏到两旁丰富多样的沙漠植物，穿过前面的花园，就可以回到阳光岛中心。

COMMERCIAL AND PUBLIC LANDSCAPE 商业公共景观

CULTURAL SQUARE 文化广场

URBAN SPACE 城市空间

| COMMERCIAL SPACE 商业空间 | OFFICE SPACE 办公空间 | HEALTHCARE AND EDUCATION 医疗教育 |

| COMMERCIAL AND PUBLIC LANDSCAPE 商业公共景观 | CULTURAL SQUARE 文化广场 | URBAN SPACE 城市空间 |

| COMMERCIAL SPACE 商业空间 | OFFICE SPACE 办公空间 | HEALTHCARE AND EDUCATION 医疗教育 |

COMMERCIAL AND PUBLIC LANDSCAPE 商业公共景观

CULTURAL SQUARE 文化广场　　URBAN SPACE 城市空间

Keywords 关键词

Material 材质
Plantation 植被
Pavement 铺装
Green Building 绿色建筑

Location: Melbourne, Australia
Client: Plenary Group, Multiplex Company, Contexx Company
Landscape Design: Aspect Studio (Australia)
Architects: Woods Bagot and NH Architecture
Bridge Design: Winward Structures

项目地点：澳大利亚墨尔本
客　　户：Plenary 集团、Multiplex 公司、Contexx 公司
景观设计：澳派（澳大利亚）景观规划设计工作室
建筑设计：Woods Bagot and NH Architecture
桥梁设计：Winward Structures

Melbourne International Convention & Exhibition Center

墨尔本国际会展中心

Features 项目亮点

Through a series of ecological measures like landscape design, floor pavement, plants selection and rainwater collection, a green building worldwide was created.

通过景观设计、地面铺装、植被选择、雨水收集等一系列生态环保措施，打造成为世界级的绿色建筑。

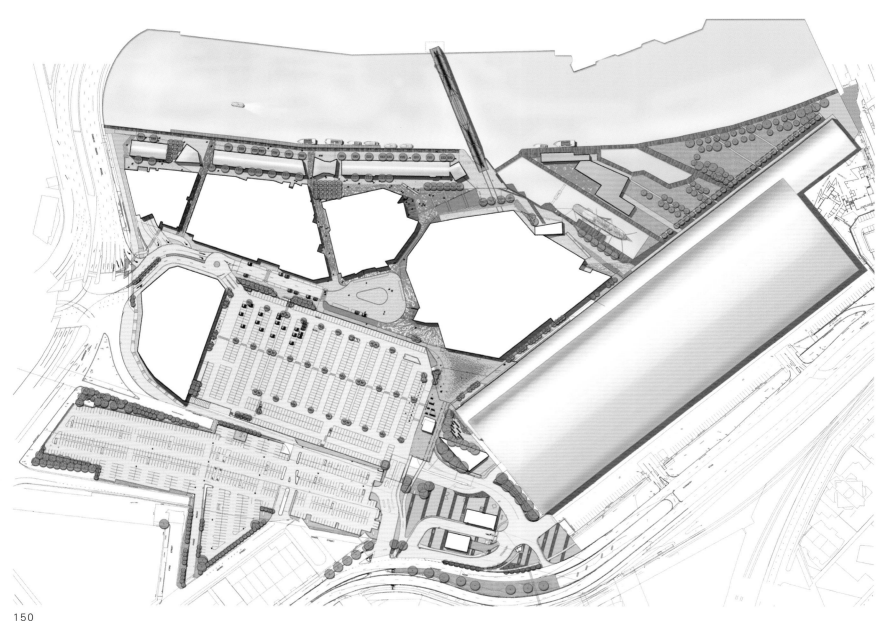

| COMMERCIAL SPACE 商业空间 | OFFICE SPACE 办公空间 | HEALTHCARE AND EDUCATION 医疗教育 |

Overview

Melbourne Convention & Exhibition Center is the first six star green building around the globe.

项目概况

墨尔本会展中心是全球第一个六星级的绿色建筑。

| COMMERCIAL AND PUBLIC LANDSCAPE 商业公共景观 | CULTURAL SQUARE 文化广场 | URBAN SPACE 城市空间 |

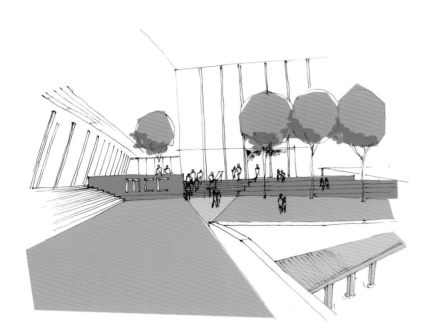
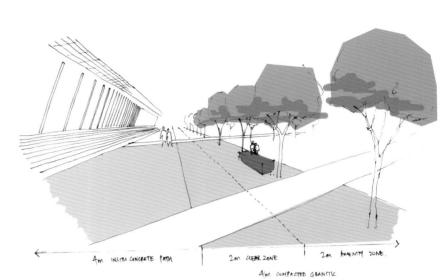

| COMMERCIAL SPACE 商业空间 | OFFICE SPACE 办公空间 | HEALTHCARE AND EDUCATION 医疗教育 |

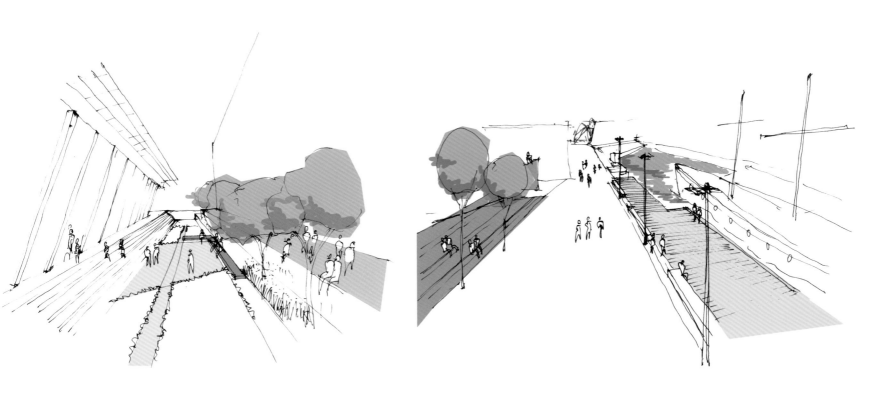

COMMERCIAL AND PUBLIC LANDSCAPE 商业公共景观

| CULTURAL SQUARE 文化广场 | URBAN SPACE 城市空间 |

| COMMERCIAL SPACE 商业空间 | OFFICE SPACE 办公空间 | HEALTHCARE AND EDUCATION 医疗教育 |

COMMERCIAL AND PUBLIC LANDSCAPE 商业公共景观

CULTURAL SQUARE 文化广场

URBAN SPACE 城市空间

Design Description

The landscape design scope in public area includes:Convention & Exhibition Center, Hilton Hotel, Commercial Retail Area, Residential Area, Public Square and Waterside Promenade.

The main principles for public space landscape design include making sure of convenient transportation 24 hours available within the site, setting up clear promenade system and cycle track system, creating Yarra riverside landscape, providing different functional facilities for residence, tourists, consumers and visitors, handling connection and passage from the center to Yarra River, utilizing and reconstructing historical landscapes.

The major pavement material is blue stone—a local material and regenerated granite and high-quality concrete pavement in the second place. Other small areas and surrounding areas utilize different pavement materials, such as pebble, granite case. Areas around trees apply infiltration floor, colorful concrete and asphalt, etc. The entire commercial area adopts tangible pavement and specify dangerous area, stairs, promenade for the convenience of blind people.

Plants selection principles are continuely using the plants on Yarra riverside avenue, plants with large crown to offer shade; plants provide shade in summer and penetrate sunshine in winter.

Rainwater on the large square is well collected to flow into the rainwater garden, filter and collect. Rainwater on the south square and parking lot is also gathered and directed into an ecological depositing pond. After the roots of plants eliminate pollutants such as engine oil and heavy metal, the water flows into Yarra River a nd Philips Port to protect the water quality.

These series of measures guarantee the fame of the first six star green building for Melbourne Convention & Exhibition Center.

| COMMERCIAL SPACE 商业空间 | OFFICE SPACE 办公空间 | HEALTHCARE AND EDUCATION 医疗教育 |

设计说明

公共区域景观设计的范围包括：会展中心、希尔顿酒店、商业零售区、住宅区、公共广场和滨水人行道。

公共空间景观设计的主要原则包括：保证地块内全天候的便捷通行，设立清晰的人行道和自行车道系统，打造雅拉河畔滨河景观，为居民、游客、购物者、参观代表等不同的人群提供不同的功能设施，处理会展中心至雅拉河的连接与通道以及历史景观的利用与重建。

公共区域路面主要的铺装材质是青石（墨尔本当地材料），其次是再生花岗岩铺地，以及高档混凝土铺地。其他小型区域和周边地区采用不同的铺地材料，如小鹅卵石、花岗岩格，乔木周围采用可渗透铺地、彩色混凝土铺地和沥青等。整个商业区使用可触知的铺地，标出危险区域、上下台阶和过街人行道，方便盲人的使用。

植被选用上延续雅拉河林阴道使用的植物类型，采用对水量吸收少、大树冠及可遮阴的植物，能够夏日遮阴，冬日透光。

大广场上地表径流的雨水得到了很好的收集，流入雨水花园进行过滤与收集。南面广场以及停车场地表径流的雨水也被导入一个生态沉淀种植池，待植物根系去除了机油、重金属等污染物后，水体才流入雅拉河与飞利浦港口，让水质得到保护。

通过这一系列的措施，保证墨尔本国际会展中心获得了全球第一个六星级绿色建筑的称呼。

COMMERCIAL AND PUBLIC LANDSCAPE 商业公共景观

CULTURAL SQUARE 文化广场

URBAN SPACE 城市空间

Keywords 关键词

Sustainability 可持续性

Green Roof 屋顶绿化

Layout 布局

Environment 环境

Location: San Francisco, CA, USA
Client: California Academy of Sciences
Landscape Design: SWA Group

项目地点：美国加州旧金山
客　　户：加州科学馆
景观设计：SWA 集团

California Academy of Sciences

加州科学馆

Features 项目亮点

The design uses scientific and technological innovation in the building and landscape and the concept of sustainability in the green roof.

绿色屋顶的设计蕴含了可持续发展的理念，科技创新被运用到了整个科学馆的建筑以及环境设计之中。

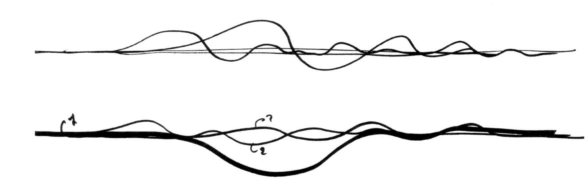

| COMMERCIAL SPACE 商业空间 | OFFICE SPACE 办公空间 | HEALTHCARE AND EDUCATION 医疗教育 |

Overview

The California Academy of Sciences, the "world's greenest museum", has earned LEED-Platinum certification through an ambitious vision for sustainable design. Landscape architecture played a major role in the realization of the design team's concept of "lifting up a piece of the park and putting a building under it", resulting in a sculptural 10 117 m² living roof that provides a unique interpretive experience and helps the project to achieve the highest level of sustainability. The roof's contours conform to the major exhibit components, research, collections and administration facilities below, and romantically echo the seven major hills of San Francisco.

项目概况

加州科学馆是世界上最"绿色"的博物馆,它已经获得了 LEED-Platinum 认证,并通过了一项雄心勃勃的设想——可持续设计,设计团队的理念是"抬起一片公园,把建筑放入其中",景观设计师在诠释这个理念上起了很大的作用。10 117 m² 如雕塑般的绿色屋顶产生了一种独特的体验,它让这个项目达到了最高级别的可持续性。屋顶外形轮廓与下面展馆内的主要展览内容、研究项目、收藏和管理设施相得益彰,巧妙地象征了旧金山的 7 座大山。

COMMERCIAL AND PUBLIC LANDSCAPE 商业公共景观

CULTURAL SQUARE 文化广场

URBAN SPACE 城市空间

| COMMERCIAL SPACE 商业空间 | OFFICE SPACE 办公空间 | HEALTHCARE AND EDUCATION 医疗教育 |

COMMERCIAL AND PUBLIC LANDSCAPE 商业公共景观

CULTURAL SQUARE 文化广场 | URBAN SPACE 城市空间

| COMMERCIAL SPACE 商业空间 | OFFICE SPACE 办公空间 | HEALTHCARE AND EDUCATION 医疗教育 |

COMMERCIAL AND PUBLIC LANDSCAPE 商业公共景观

CULTURAL SQUARE 文化广场　　URBAN SPACE 城市空间

Keywords 关键词

Central Space 中央空间

Transformation 改造重生

Place 场所感

Material 材料

Location: Providence, RI, USA
Design: Klopfer Martin Design Group

项目地点：美国罗德岛州普罗维登斯
设计单位：克洛普弗·马丁公司

The Steel Yard
钢铁工厂院落

Features 项目亮点

Using colorful carpet pavement for the factory outer space, the design activates every small space functionally and creatively.

运用彩色的地毯式的铺装将工厂的外部空间进行整合设计，从而激活了场所内的各个小空间的功能与创意。

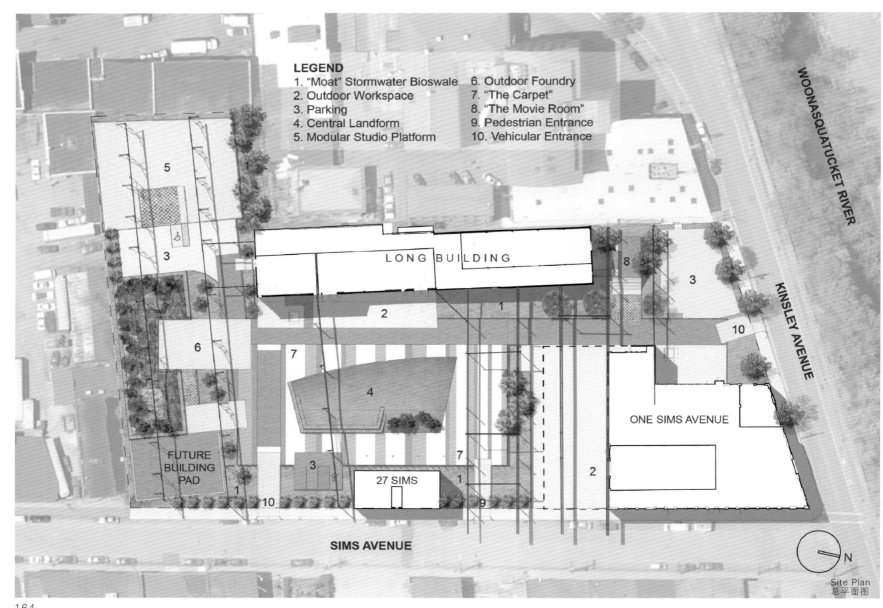

Site Plan 总平面图

| COMMERCIAL SPACE 商业空间 | OFFICE SPACE 办公空间 | HEALTHCARE AND EDUCATION 医疗教育 |

Design Description

Required exterior spaces include a primary central space (fashioned as a multi-colored paved "carpet") that allows for individual and group work, staging of large events with audiences of up to several hundred, car rallies, farmer's markets, etc., and whose character defines a sense of place. This is surrounded by secondary work spaces such as interior/exterior spill-out shop spaces, an outdoor foundry, a "hang-out" space for movie nights and relaxation, and a future visiting artist's studio. Tertiary service spaces include storage for raw materials and finished art pieces, a paved space serving incubator businesses and artists in shipping container studios, and 20 parking spaces.

设计说明

项目改造需要的外部空间包括一个主要的中央空间（采用了彩色"地毯"式的铺装），可以用于个人或集体活动，承办容纳几百人的大型活动，公路赛车、农贸市场等，它的特点给予了这个地方很强的场所感。这个中央空间被第二层次的工作区域空间环绕，它包括内部和外部突出的商店空间、户外的铸造厂、举行电影之夜和放松的好去处，以及未来可供参观的艺术家工作室等。第三个层次是产业服务空间，包括原材料和已完工的艺术品的储存空间，以及将一块铺设好的空间设计成集装箱工作室来服务于创意产业和艺术家，还包括20个停车位。

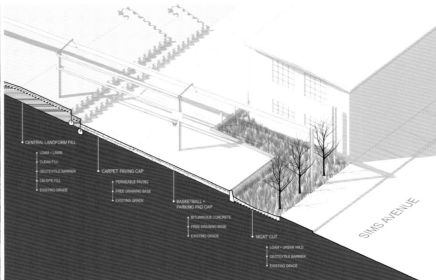

COMMERCIAL AND PUBLIC LANDSCAPE 商业公共景观

CULTURAL SQUARE 文化广场

URBAN SPACE 城市空间

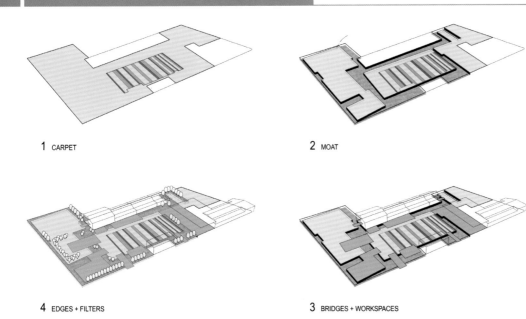

1 CARPET

2 MOAT

4 EDGES + FILTERS

3 BRIDGES + WORKSPACES

ORGANIZATIONAL STRATEGY

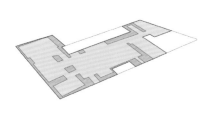

1 TOTAL CAP SURFACE

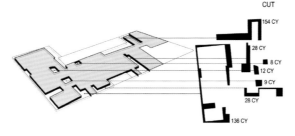

2 EXTRACTION (MOAT)

4 VEGETATION (URBAN WILD + TURF)

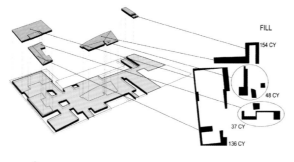

3 AGGREGATION (LANDFORM)

REMEDIATION STRATEGY

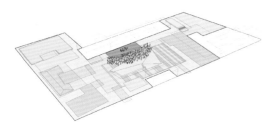

IRON CHEF

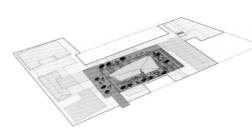

CRUISE NIGHT

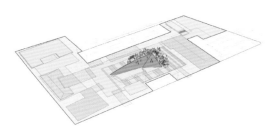

HALLOWEEN IRON POUR

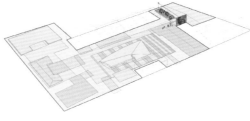

MOVIE NIGHT

EVENT DIAGRAMS

| COMMERCIAL SPACE 商业空间 | OFFICE SPACE 办公空间 | HEALTHCARE AND EDUCATION 医疗教育 |

COMMERCIAL AND PUBLIC LANDSCAPE 商业公共景观

CULTURAL SQUARE 文化广场

URBAN SPACE 城市空间

| COMMERCIAL SPACE 商业空间 | OFFICE SPACE 办公空间 | HEALTHCARE AND EDUCATION 医疗教育 |

| COMMERCIAL AND PUBLIC LANDSCAPE 商业公共景观 | CULTURAL SQUARE 文化广场 | URBAN SPACE 城市空间 |

Keywords 关键词

Social Space 社交空间
Abstract Style 抽象风格
Modern Elements 现代元素
Rainwater Utilization 雨水利用

Location: San Francisco, CA, USA
Landscape Design: OLIN

项目地点：美国加州旧金山
景观设计：OLIN

Gardens of Corporate Headquarters

公司总部花园

Features 项目亮点

Skillfully mixing two different spatial structures with modern techniques, the design cleverly uses rainwater to mitigate the adverse effects of the surrounding environment to create a new social space.

设计大胆地将两个不同的空间结构用现代的手法巧妙糅为一体，创造了一个新的社交空间，并很好地利用了雨水来减轻对周边环境的不利影响。

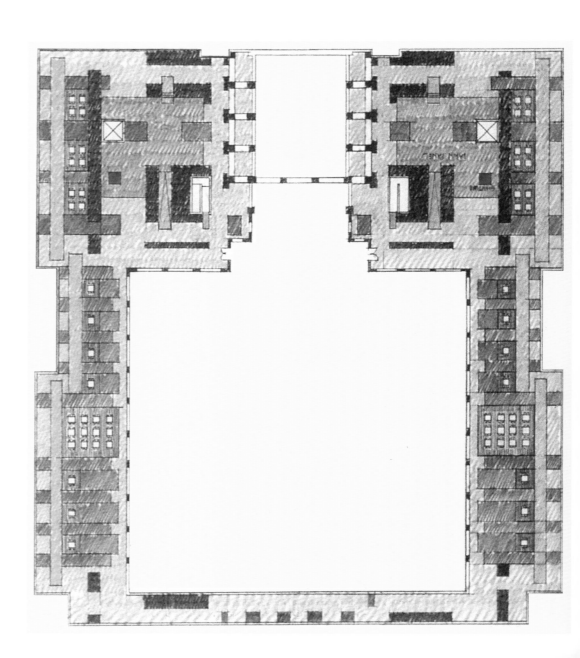

| COMMERCIAL SPACE 商业空间 | OFFICE SPACE 办公空间 | HEALTHCARE AND EDUCATION 医疗教育 |

COMMERCIAL AND PUBLIC LANDSCAPE 商业公共景观

| CULTURAL SQUARE 文化广场 | URBAN SPACE 城市空间 |

Design Description

Modern, fresh and clean, the gardens of this corporate headquarters are inspired by the abstract works of the De Stijl movement. Comprised of two distinct spaces, both over structure, the design incorporates fine materiality into bold, modern gestures that reflect the corporation's identity as well as its key location on the thriving waterfront of San Francisco. Over a former industrial site, the gardens create new social space and assist in managing stormwater and mitigating the urban heat island effect.

设计说明

该公司总部花园设计受 De Stijl 抽象作品风格启发，总体现代、清新、干净。它由两个完全分开的部分组成，每个部分都在建筑上。景观的设计是用精细的材料来体现大胆的现代风格，这个风格能反映公司的特性。项目处于旧金山繁华的滨水区，凸显出该公司重要的社会地位。项目之前是一块工业用地，花园创造出一个新的社交空间，并且通过有效利用雨水，减轻城市热岛效应。

| COMMERCIAL SPACE 商业空间 | OFFICE SPACE 办公空间 | HEALTHCARE AND EDUCATION 医疗教育 |

COMMERCIAL AND PUBLIC LANDSCAPE 商业公共景观

| CULTURAL SQUARE 文化广场 | URBAN SPACE 城市空间 |

| COMMERCIAL SPACE 商业空间 | OFFICE SPACE 办公空间 | HEALTHCARE AND EDUCATION 医疗教育 |

COMMERCIAL AND PUBLIC LANDSCAPE 商业公共景观

CULTURAL SQUARE 文化广场

URBAN SPACE 城市空间

Keywords 关键词

Bio-diversity 生态多样化

Designed for Comfort 舒适性设计

Rainwater Systems 雨洪系统

Green Roofs 绿色屋顶

Location: McLean, Virginia, USA
Client: Gannett, Inc.
Landscape Design: Michael Vergason Landscape Architects, Ltd.

项目地点：美国维吉利亚州麦克莱恩市
项目委托：Gannett 集团
设计单位：Michael Vergason 景观设计事务所

Gannett—USA Today Headquarters
Gannett——《今日美国》总部大楼

Features 项目亮点

The landscape architect creates an ecologically diverse refuge by seamlessly weaving the indoor and outdoor spaces into a campus of the green roof and a rich palette of trees.

设计师将室内外空间与周围环境完美结合，设计了绿色屋顶，种植了色彩丰富的树木，创造出一个生态多样化的圣地。

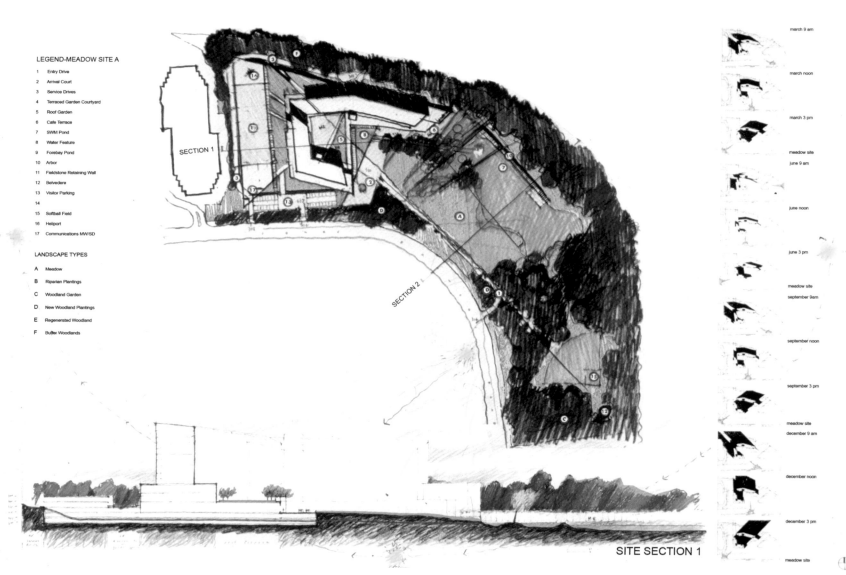

MEADOW SITE A

| COMMERCIAL SPACE 商业空间 | OFFICE SPACE 办公空间 | HEALTHCARE AND EDUCATION 医疗教育 |

COMMERCIAL AND PUBLIC LANDSCAPE 商业公共景观

| CULTURAL SQUARE 文化广场 | URBAN SPACE 城市空间 |

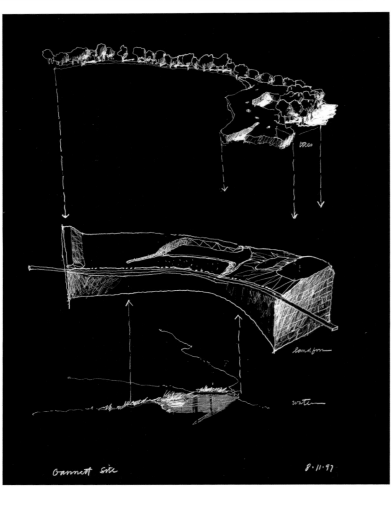

Overview

The USA Today Headquarters is an ecologically diverse refuge in the rapidly developing business and retail center of Tysons Corner, Virginia. The landscape architect developed a site strategy that seamlessly weaves the indoor and outdoor spaces into a campus of extensive roof gardens and terraces, riparian plantings and preserved woodlands.

项目概况

对商业和零售发展快速的维吉利亚州泰森斯科纳地区来说,《今日美国》总部大楼的景观设计无疑是一个生态多样化的圣地。景观设计师的设计目标就是将室内外的空间与周围广阔的屋顶花园、倾斜平地、水渠边植物和保护林地天衣无缝地结合起来。

| COMMERCIAL SPACE 商业空间 | OFFICE SPACE 办公空间 | HEALTHCARE AND EDUCATION 医疗教育 |

| COMMERCIAL AND PUBLIC LANDSCAPE 商业公共景观 | CULTURAL SQUARE 文化广场 | URBAN SPACE 城市空间 |

Design Description

The landscape architect tells a story about the site, integrating the existing stormwater system and creating interplay with the green roof, a rich palette of trees at varying levels, and other amenities. A new approach to corporate typology that feels more natural and organic. This project demonstrates the exceptional value of thoughtful site design and site repair for the creation of a distinctive corporate campus.

设计说明

景观设计师将现有雨洪系统整合起来，创造了与之相互影响的绿色屋顶，种植了不同季节色彩丰富的树木，并进行了其他舒适性设计。设计体现了类型学的新理念，更加自然和有机。这个工程的设计为以后类似的设计作出了很好的榜样。

COMMERCIAL AND PUBLIC LANDSCAPE 商业公共景观

CULTURAL SQUARE 文化广场

URBAN SPACE 城市空间

| COMMERCIAL SPACE 商业空间 | OFFICE SPACE 办公空间 | HEALTHCARE AND EDUCATION 医疗教育 |

Healthcare and Education
医疗教育

Integrated Design
Identity System
Participation
Eco and Harmony

整合设计
标识系统
参与性
生态和谐

COMMERCIAL AND PUBLIC LANDSCAPE 商业公共景观

CULTURAL SQUARE 文化广场

URBAN SPACE 城市空间

Keywords 关键词

Transform 改建

Materials 材料

Campus Space 校园空间

Stormwater Solution 雨水解决方案

Location: Mesa, AZ, USA
Client: Arizona State University
Architects: Ten Eyck Landscape Architects, Inc., Phoenix

项目地点：美国亚利桑那州梅沙地区
客　户：亚利桑那州立大学
设　计：Ten Eyck Landscape Architects, Inc., Phoenix

Arizona State University Polytechnic Campus— New Academic Complex

亚利桑那州立大学理工学院——新教学综合大楼

Features 项目亮点

The transformed arroyo and drainage system has been set as stormwater solution with high efficiency to provide the irrigation, lessen the effect on flood prone area and create a comfortable space for the campus.

利用改建的小河及校园排水系统，建立高效的雨水解决方案，既能灌溉、减少雨水对洪水易发区的影响，又为校园创造了一方舒适空间。

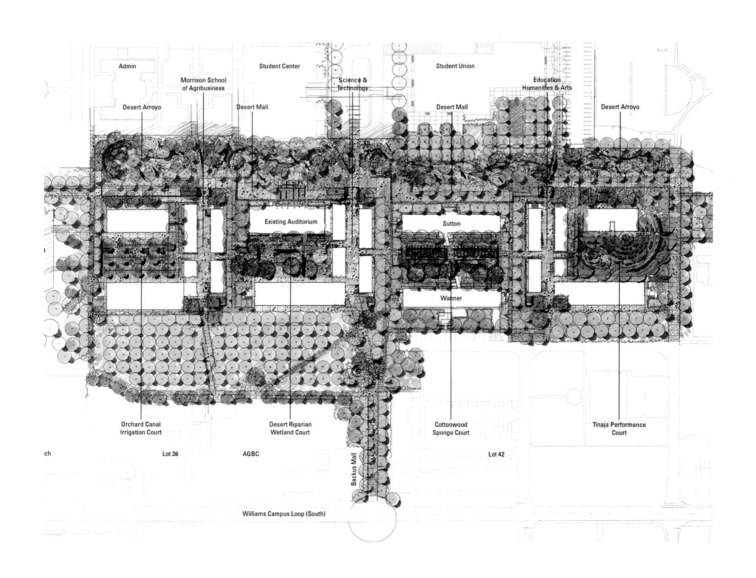

| COMMERCIAL SPACE 商业空间 | OFFICE SPACE 办公空间 | **HEALTHCARE AND EDUCATION 医疗教育** |

Overview

The Arizona State University (ASU) Polytechnic Campus project consists of 21 acres of site work in association with five new classroom building complexes. The goal was to transform the barren site of the former Air Force base into a thriving Sonoran Desert campus for learning.

项目概况

美国亚利桑那州立大学（ASU）理工学院项目包括 5 栋全新的综合教学楼，共占地 21 英亩（约 84 983 m²）。项目的目标是将一个废弃的空军基地改造成索诺兰沙漠中的一个校园。

COMMERCIAL AND PUBLIC LANDSCAPE 商业公共景观 | CULTURAL SQUARE 文化广场 | URBAN SPACE 城市空间

Design Description

A major large asphalt street that formerly flooded during rains was transformed into a permeable, water harvesting arroyo adjacent to new campus malls and courtyards, giving students and faculty a daily connection to nature and each other while celebrating the path of precious, ephemeral rain water on campus.

Inspired by the client's need for stormwater solutions and the Sonoran Desert's arroyos, the design of the desert mall allows campus drainage to meander through a new, high performance, water cleansing native landscape adjacent to new major east west pedestrian circulation through the campus. Stabilized decomposed granite walkways in combination with desert shade trees help to reduce the urban heat island effect while the judicious use of textured concrete paving ensures proper emergency access required for a campus setting. Salvaged desert trees and small native shrubs, cacti and seed were used to create the arroyo habitat. During a rain event the arroyo captures and slows the stormwater providing supplemental deep watering to the vegetation while lessening the effect of run-off to downstream offsite properties by retaining water within this historically troublesome, flood prone area.

Situated on the former Williams Air Force Base, the existing conditions consisted of acres of asphalt, extensive concrete sidewalks and expanses of river rock covered ground. These materials were re-used for new outdoor spaces and parking areas. Harvested concrete paving was transformed into campus seating elements and retaining walls throughout the project. Existing river rock was used in gabions to construct both freestanding and retaining walls to create another layer of enclosure to the courtyards. Drought tolerant native plants were woven together with judicious hardscape to create comfortable, cooler spaces throughout to encourage social interaction and human comfort. The landscape nourished by stormwater also provides habitat for desert fauna that visit the site. A state of the art drip irrigation system was used to allow many of the zones to eventually be turned off completely of water.

| COMMERCIAL SPACE 商业空间 | OFFICE SPACE 办公空间 | HEALTHCARE AND EDUCATION 医疗教育 |

设计说明

设计师们将一条现有的柏油马路改建为具有一定渗透性和储水功能的小河，流过新建的大楼和校园，使全校师生在日常生活中也能亲近自然，增加了人与人之间的沟通机会，同时也解决了之前柏油马路遇到雨天经常被淹没的难题。

客户提出了雨水解决方案和索诺兰沙漠中小河的需求，受此启发设计师们设计出了沙漠购物中心，让校园排水系统以蜿蜒的姿态穿过新的高性能水质净化地貌，毗邻东西方向的主人行道。花岗岩人行道与沙漠遮阴树有助于减少城市热岛效应，带有纹理的混凝土路面明智地确保了适当的紧急通道。回收的沙漠树和小原生灌木、仙人掌及种子被用来创建小河这一栖息地。下雨时小河能够收集雨水并减缓其流动速度，为植被补充深浇水，减少径流对下游历史洪水易发区的影响。

前威廉姆斯空军基地现存数英亩的沥青、混凝土人行道，及用大量河石铺就的路面。这些材料被重新用于新的室外空间和停车区。大量的混凝土被改造成校园座椅和挡土墙，河石被制作成石笼，用于构建挡土墙，为各个空间提供多一层的外壳。耐旱原生植物与硬质景观交织在一起，创造舒适凉爽的空间，提升社会互动和人体舒适度。用雨水滋润的景观也为沙漠动物提供了栖息地，许多区域都采用了国家最先进的滴灌系统，减少了自来水的使用。

| COMMERCIAL AND PUBLIC LANDSCAPE 商业公共景观 | CULTURAL SQUARE 文化广场 | URBAN SPACE 城市空间 |

| COMMERCIAL SPACE 商业空间 | OFFICE SPACE 办公空间 | HEALTHCARE AND EDUCATION 医疗教育 |

COMMERCIAL AND PUBLIC LANDSCAPE 商业公共景观 | CULTURAL SQUARE 文化广场 | URBAN SPACE 城市空间

| COMMERCIAL SPACE 商业空间 | OFFICE SPACE 办公空间 | HEALTHCARE AND EDUCATION 医疗教育 |

COMMERCIAL AND PUBLIC LANDSCAPE 商业公共景观

CULTURAL SQUARE 文化广场　　URBAN SPACE 城市空间

Keywords 关键词
Ecology 生态
Master Plan 规划
Green Space 绿色空间
Natural Environment 自然环境

Location: Kashiwa-shi, Chiba Prefecture, Japan
Client: The Hiroike Institute of Education
Landscape Design: KEIKAN SEKKEI Tokyo Co., Ltd.
Area: 8,000 m²

项目地点：日本千叶县柏市
客　　户：广池教育学院
景观设计：日本东京景观设计株式会社
面　　积：8 000 m²

Reitaku University—New School Building in the Woods

日本丽泽大学——树丛中的新教学楼

Features 项目亮点

The architects combine the campus and its surrounding natural landscape environment by reasonable planning, create nature closed, ecological and harmonious green environment.

设计师通过合理规划，将校园空间融入周边自然景观环境，打造出亲近自然、生态和谐的绿色空间。

| COMMERCIAL AND PUBLIC LANDSCAPE 商业公共景观 | CULTURAL SQUARE 文化广场 | URBAN SPACE 城市空间 |

Overview

Reitaku University and its urban forest have evolved together, resulting in a well defined campus set in a mature landscape.

项目概况

丽泽大学校园已延伸至周边树林，校园与周边自然环境形成美不胜收的自然景观。

| COMMERCIAL SPACE 商业空间 | OFFICE SPACE 办公空间 | HEALTHCARE AND EDUCATION 医疗教育 |

| COMMERCIAL AND PUBLIC LANDSCAPE 商业公共景观 | CULTURAL SQUARE 文化广场 | URBAN SPACE 城市空间 |

| COMMERCIAL SPACE 商业空间 | OFFICE SPACE 办公空间 | HEALTHCARE AND EDUCATION 医疗教育 |

| COMMERCIAL AND PUBLIC LANDSCAPE 商业公共景观 | CULTURAL SQUARE 文化广场 | URBAN SPACE 城市空间 |

| COMMERCIAL SPACE 商业空间 | OFFICE SPACE 办公空间 | HEALTHCARE AND EDUCATION 医疗教育 |

| COMMERCIAL AND PUBLIC LANDSCAPE 商业公共景观 | CULTURAL SQUARE 文化广场 | URBAN SPACE 城市空间 |

| COMMERCIAL SPACE 商业空间 | OFFICE SPACE 办公空间 | HEALTHCARE AND EDUCATION 医疗教育 |

| COMMERCIAL AND PUBLIC LANDSCAPE 商业公共景观 | CULTURAL SQUARE 文化广场 | URBAN SPACE 城市空间 |

Design Description

Informed by analysis of campus wide open space, a Green Space Zoning Plan was developed based on the concept of "Satoyama" or undeveloped rural nature areas which encourage villagers to maintain contact with the natural environment. This proposal established a framework, or "Skeleton Green" which organized the various sub-component landscape spaces, termed "Infill Greens".

In an educational facility that literally shares a single breath with the surrounding forest, architecture and landscape integrated into a single learning environment promoting coexistence through: the preservation of the woods, the creation of the woods, and protection of the woods. This project demonstrates a strategy for harmonizing new construction in the context of the larger campus plan maintaining the integrity of the urban forest and ecological corridors while providing opportunities for "Infill Greens" capable of adapting with the changing needs of the students and university.

| COMMERCIAL SPACE 商业空间 | OFFICE SPACE 办公空间 | HEALTHCARE AND EDUCATION 医疗教育 |

设计说明

通过分析校园内广阔的开放空间，设计师在"里山"的基础上打造了一片绿色空间地带，或者说是未开发的原始地带以鼓励人们多亲近自然。该项目打造了一个绿色框架，合理规划了景观空间。

环绕着校园的森林与校园景观完美融合，建筑与景观融合的学习环境的共存性主要体现在：保存教学楼周边的大树，并种植更多树木。该项目体现了新建筑与大学校园规划统一的策略。保持城市森林与生态绿道的融合，同时提供绿化空间以适应学校和学生的需求。

COMMERCIAL AND PUBLIC LANDSCAPE 商业公共景观

CULTURAL SQUARE 文化广场　　URBAN SPACE 城市空间

Keywords 关键词
- Green 绿色
- Space 空间
- Ecology 生态
- Environment 环境

Location: Sendai, Japan
Landscape Design: Keikan Sekkei Tokyo Co., Ltd.

项目地点：日本仙台
景观设计：日本东京景观设计株式会社

Campus Mall at Tohoku Pharmaceutical University
东北药科大学广场

Features 项目亮点

Take advantage of topographical change, carefully select plants, create open landscape view, the campus integrates with the ecological environment, harmonious and comfortable.

充分利用广场地形地势的变化，精选植栽，营造开阔景观视野，将校园融入绿色生态环境，和谐舒适。

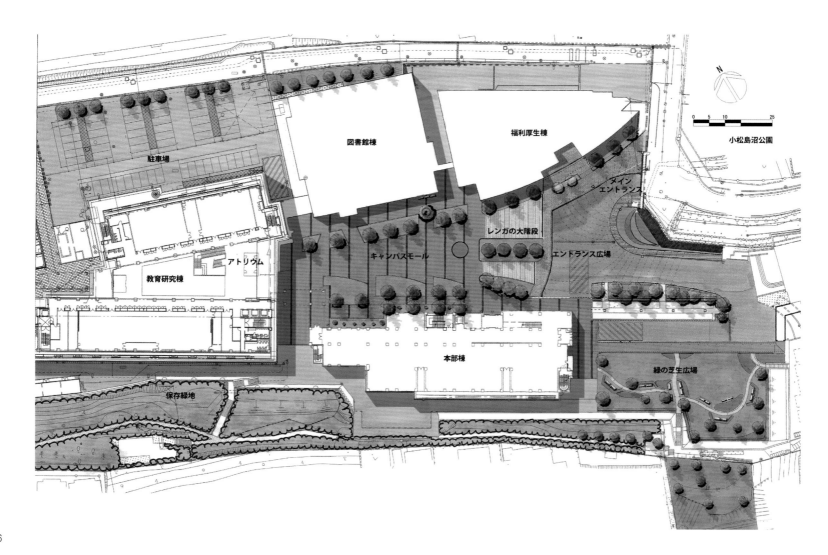

| COMMERCIAL SPACE 商业空间 | OFFICE SPACE 办公空间 | HEALTHCARE AND EDUCATION 医疗教育 |

COMMERCIAL AND PUBLIC LANDSCAPE 商业公共景观

CULTURAL SQUARE 文化广场

URBAN SPACE 城市空间

Overview

Adjacent to Komatsushima Pond Park and surrounded by a series of open green spaces, the campus is a critical part of the larger green network.

项目概况

该项目毗邻小松岛湖公园，周围环绕着开放绿色空间，东北药科大学为大范围绿色网络的关键部分。

| COMMERCIAL SPACE 商业空间 | OFFICE SPACE 办公空间 | HEALTHCARE AND EDUCATION 医疗教育 |

Design Description

Highlighting the sites environmental characteristics, including significant topographical change, and a desire for the campus to perform as part of the larger natural system became the vital goals for this project. Taking advantage of the major topographical change on site, a grand brick staircase was designed as a focal point at the main entrance, creating a dynamic approach to the campus. Campus buildings are located along the central "Campus Mall" which opens onto Komatsushima Pond, providing continuous open space and visual depth for the campus, physically and visually linking it to the larger green network. The "Campus Mall" provides students and faculty opportunities for outdoor relaxation, also functioning as critical pedestrian circulation. Careful selection of plant material and trees sized to balance with those existing in the preserved forest integrate the two spaces bringing a forest grove into the heart of the campus. The use of brick pavers consistent with architectural materials unifies the architecture and landscape, creating a harmonious space for campus life.

设计说明

设计注重场地环境特征，包括显著的地形变化，旨在将校园变成自然系统的一部分。利用地势起伏的特点在进口处用砖砌成了阶梯，使通往校园的路变得动感。校园内的建筑沿着中心广场分布，面向小松岛湖，使校园空间和视野变得开阔，并融入绿色的生态环境。校园广场为全校师生提供室外放松地，也作为主要人行通道。新栽种的树木通过精心挑选维持了与原树林的平衡性并将周边的树林延伸至校园内。建筑材料与景观一致打造了和谐的校园生活空间。

| COMMERCIAL SPACE 商业空间 | OFFICE SPACE 办公空间 | HEALTHCARE AND EDUCATION 医疗教育 |

COMMERCIAL AND PUBLIC LANDSCAPE 商业公共景观

CULTURAL SQUARE 文化广场

URBAN SPACE 城市空间

| COMMERCIAL SPACE 商业空间 | OFFICE SPACE 办公空间 | HEALTHCARE AND EDUCATION 医疗教育 |

COMMERCIAL AND PUBLIC LANDSCAPE 商业公共景观

| CULTURAL SQUARE 文化广场 | URBAN SPACE 城市空间 |

| COMMERCIAL SPACE 商业空间 | OFFICE SPACE 办公空间 | HEALTHCARE AND EDUCATION 医疗教育 |

COMMERCIAL AND PUBLIC LANDSCAPE 商业公共景观

CULTURAL SQUARE 文化广场 | URBAN SPACE 城市空间

Keywords 关键词
- Green Space 绿色环境
- Lighting Design 灯光设计
- Orientation Point 指示方位
- Healing Environment 氛围营造

Location: Rotterdam, The Netherlands
Landscape Design: Stijlgroep landscape and urban design
Copyright: Stijlgroep landscape and urban design (plans/ text)
Samson urban elements, The Netherlands (photos)

项目地点：荷兰鹿特丹
景观设计：Stijlgroep 景观和城市设计事务所
版权所有：Stijlgroep 景观和城市设计事务所（平面图 / 文字）
荷兰 Samson urban elements（照片）

Maastad Hospital, Rotterdam
鹿特丹玛莎塔德医院

Features 项目亮点

With special lighting design, the patios themselves create a quiet and secure oasis for patients, visitors and staff.

通过独特的灯光设计，医院天井为患者、访客和医护工作者提供了一方宁静祥和的绿洲。

| COMMERCIAL SPACE 商业空间 | OFFICE SPACE 办公空间 | HEALTHCARE AND EDUCATION 医疗教育 |

Overview

Getting lost in a huge building is not only a fear of most users, but happens quite regular. The new hospital of Rotterdam South is one of those huge developments with an enormous building mass. The design team came up with a unique concept for the five patios being located along the central internal corridor with a length of 500 meters.

项目概况

对于多数使用者来说，在巨型建筑中迷路是件令人担忧且十分常见的事。位于鹿特丹南部的这座新医院便是这样一座体量庞大的建筑。设计师对医院中央走廊附近的五处天井进行了改建，总长度约 500 m。

COMMERCIAL AND PUBLIC LANDSCAPE 商业公共景观

| CULTURAL SQUARE 文化广场 | URBAN SPACE 城市空间 |

Design Description

Sustainability means for designers to focus on the need of the users. With the concept of three different themes combined with special lighting elements the right answer in this context was given: The patios are as orientation point the central heart of the new development and meet the demands as part of the healing environment. Special lighting elements emphasize the identity of each patio. In the first two patios massive rectangular lighting columns accomplish a minimalistic design whereas oversized seating elements, which are lit during evening and at night time, are the main element of the central patio. Those seating elements dare users to sit under the soft shade of the huge honey locust trees. A modern interpretation of a bamboo garden has been chosen as topic for the last pair of gardens. A diverse mix of bamboo and grasses create the perfect scenery for the gigantic bamboo lighting.

What seems to be very special already at daytime is brought to life in the evening and at night time. Soft, sustainable LED light changing colors and shades attract attention and dare to explore, to experience the gardens also after sunset. A second layer of lighting has been added in every patio: from midnight on the feature lighting is replaced by bollard lighting to reduce the intensity of light.

A dream became reality, a dream to get lost in paradise: the patios of the hospital. The patios themselves create a quiet and secure oasis for patients, visitors and staff. Additional advantage of these green patios is the positive affect on recuperating patients. Green spaces have simply a very positive influence on people referring to the concept of a healing environment.

| COMMERCIAL SPACE 商业空间 | OFFICE SPACE 办公空间 | HEALTHCARE AND EDUCATION 医疗教育 |

设计说明

设计充分考虑了使用者的需求，选择了三个不同的主题，运用特殊的灯光效果营造出不同的气氛。天井不仅指示了医院的中心位置，并且与周围环境相融，增强了医院的治愈氛围。每个天井采用不同的灯光元素，突出其个性。在头两个天井中，设有巨大的矩形灯柱和宽大的座椅，还种植有高大的皂荚树，成为了这里的中心区域。后两个天井的装饰主题则是竹林，这里种植的是竹子和各种草本植物，与巨大的竹子形状照明灯相呼应。

当夜幕降临时，独特的灯光设计将白天的奇特景观也同样呈现出来。所有天井里都有环保的 LED 灯，不断变化的色调能营造出独特的气氛，吸引人们前往探索、体验花园日落后的景观。而到了午夜，天井则改用灯柱照明，减弱了灯光的强度，烘托出静夜的温馨氛围。

通过奇妙的灯光设计，医院的天井实现了人们置身天堂的梦想。天井为患者、访客和医护工作者提供了一方宁静祥和的绿洲。同时，因为绿色的环境有助于人们的康复治疗，所以绿色天井还给康复中的患者以积极的影响。

COMMERCIAL AND PUBLIC LANDSCAPE 商业公共景观

| CULTURAL SQUARE 文化广场 | URBAN SPACE 城市空间 |

| COMMERCIAL SPACE 商业空间 | OFFICE SPACE 办公空间 | HEALTHCARE AND EDUCATION 医疗教育 |

COMMERCIAL AND PUBLIC LANDSCAPE 商业公共景观

| CULTURAL SQUARE 文化广场 | URBAN SPACE 城市空间 |

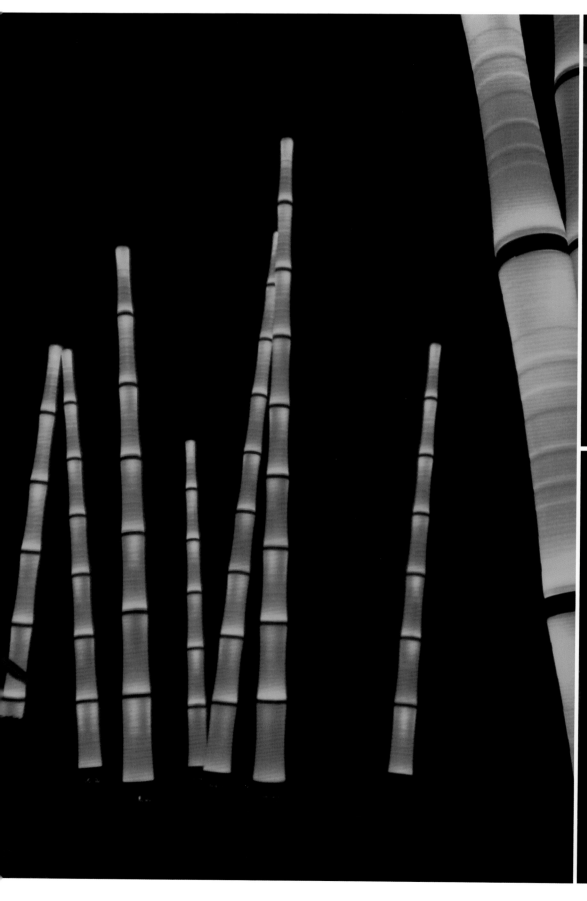

| COMMERCIAL SPACE 商业空间 | OFFICE SPACE 办公空间 | HEALTHCARE AND EDUCATION 医疗教育 |

| COMMERCIAL AND PUBLIC LANDSCAPE 商业公共景观 | CULTURAL SQUARE 文化广场 | URBAN SPACE 城市空间 |

Keywords 关键词

Landscape Space 景观空间
Integrated Design 整合设计
Vegetation 植被
Environment 环境

Location: Johannesburg, South Africa
Landscape Design: GREENinc Landscape Architecture
Architects: KCS Architects

项目地点：南非约翰内斯堡
景观设计：南非 GREENinc 景观设计事务所
建筑设计：KCS 建筑设计事务所

Johannesburg City Park Environmental Education & Research Centre

约翰内斯堡城市公园环境教育和研究中心

Features 项目亮点

It combines the landscape design tightly with the surroundings, and shows the originality and naturality in the vegetation selection.

将景观设计与周围的环境紧密的结合，同时在植被的选择上突出本土性与自然性。

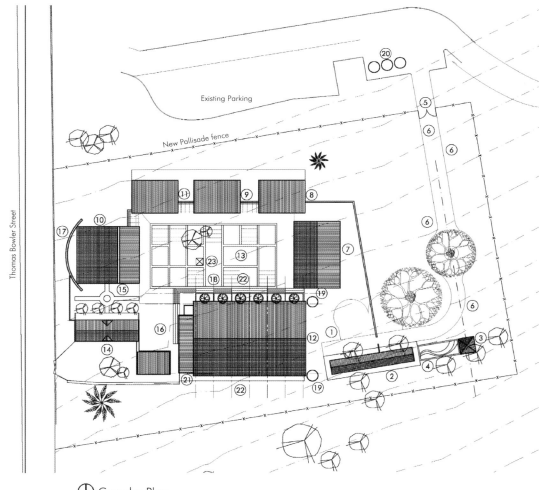

Complex Plan

1. Arrival court
2. Tuck shop + Gifts
3. Look out tower + Information
4. Reflective pond
5. Main gate
6. Entrance route + gardens
7. Museum
8. Library
9. Herberium + Laboratory
10. Auditorium
11. Offices
12. Ineractive shed with pods
13. Terraced gardens
14. Overnight accommodation
15. Formal garden
16. Grey water court.
17. Signage wall
18. Tea Garden
19. Rainwater tank
20. Recycle bins
21. Toilets
22. Children's play area
23. Windmill

⊕ Complex Plan

| COMMERCIAL SPACE 商业空间 | OFFICE SPACE 办公空间 | HEALTHCARE AND EDUCATION 医疗教育 |

Overview

From the outset of the project, the architects decided that despite the financial constraints and the absence of a clear brief, it was their responsibility to approach the project in an environmentally sensitive manner. The aim was to create an inviting, environmentally responsive landscape, for showcasing indigenous vegetation and providing opportunities for gathering and learning.

项目概况

尽管受到资金的限制，也还没有一个明确的设计理念，但设计师从一开始就认为设计必须本着对环境负责的态度。项目目标是创造一个引人注目的融于环境的景观设计，展示当地的植被特色，可作为收集和学习的案例。

COMMERCIAL AND PUBLIC LANDSCAPE 商业公共景观 | CULTURAL SQUARE 文化广场 | URBAN SPACE 城市空间

Design Description

The existing formal garden (with its abundance of exotic trees and planting) at the entrance to the complex, posed a paradoxical design challenge. In theory an environmental education center should showcase the best of what the local natural environment has to offer and not exotic vegetation. To keep true to the ideals of an environmental education center, the architects decided to juxtapose the exotic landscape of the past with the indigenous landscape of the development. The transition or threshold between these two ideologies would be physically represented by a low dividing wall.

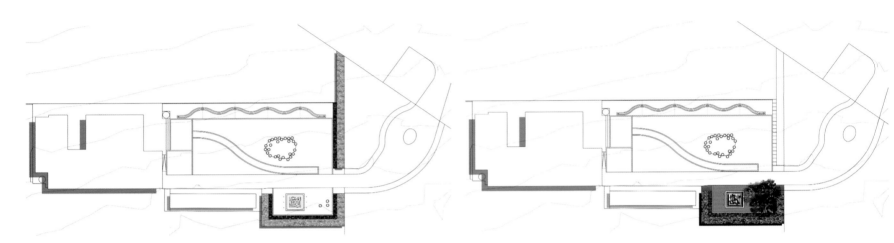

| COMMERCIAL SPACE 商业空间 | OFFICE SPACE 办公空间 | HEALTHCARE AND EDUCATION 医疗教育 |

设计说明

项目入口处有一个现存的花园（拥有丰富的外来树木和植物），对整体设计构成了一个矛盾型挑战。在理论上，环境教育中心应展现当地自然环境所提供的最好的东西，而不是这些外来的植物。为了突出环境教育中心的实质，设计师决定把过去的外来植物一同并入本土景观建设中。这两种意识形态之间的过渡从一个较低的分隔墙中便可体现出来。

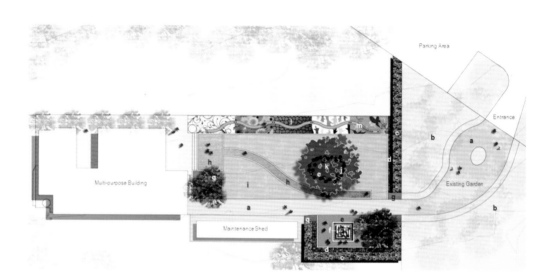

COMMERCIAL AND PUBLIC LANDSCAPE 商业公共景观

CULTURAL SQUARE 文化广场

URBAN SPACE 城市空间

COMMERCIAL AND PUBLIC LANDSCAPE 商业公共景观

CULTURAL SQUARE 文化广场 | URBAN SPACE 城市空间

Keywords 关键词

- Art 艺术特性
- Lawn Markings 草坪空间
- Text Inscriptions 铭文雕刻
- Material 材料

Location: Johannesburg, South Africa
Client: University of Johannesburg
Landscape Design: GREENinc Landscape Architects
Architects: ARC & Mashabane Rose Architects, in joint venture

项目地点：南非约翰内斯堡
客　　户：约翰内斯堡大学
景观设计：南非 GREENinc 景观设计事务所
建筑设计：ARC & Mashabane Rose Architects 合资企业

University of Johannesburg Arts Center
约翰内斯堡大学艺术中心

Features 项目亮点

The perfect integration of the building and its surrounding gives the project strong artistic quality, and the crafty space layout creates the harmonious atmosphere among the architecture, landscape and people.

建筑与场地环境的完美融合催生了该设计具有极强的艺术性，巧妙的空间布局关系则很好地实现了建筑、景观与人之间的和谐氛围。

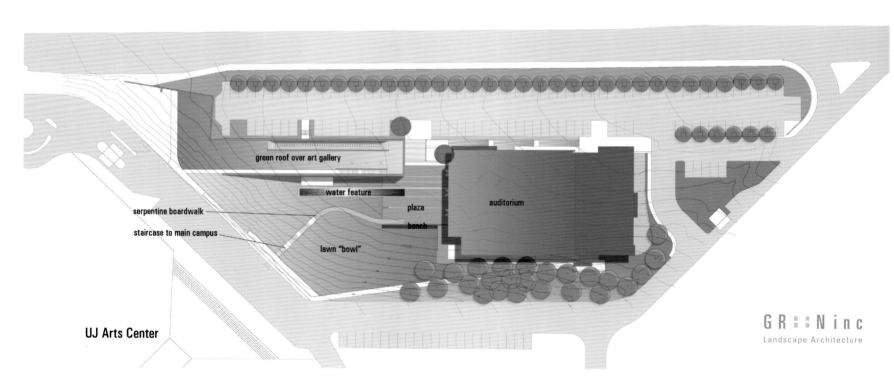

Overview

The project integrates the new buildings of the complex into the surrounding landscape of the campus and links them with the existing buildings. Landscape design and architecture are seamlessly melded to the extent that it is difficult to imagine one working without the other.

项目概况

该项目把新建筑与周围的校园景观结合起来,并与校园里现存的建筑联系在一起。景观设计和建筑设计紧密地融合在一起,两者缺一不可。

COMMERCIAL AND PUBLIC LANDSCAPE 商业公共景观

CULTURAL SQUARE 文化广场

URBAN SPACE 城市空间

| COMMERCIAL SPACE 商业空间 | OFFICE SPACE 办公空间 | HEALTHCARE AND EDUCATION 医疗教育 |

Design Description

The landscape also integrates art into its fabric in various ways from poetry to lawn markings for which the clean sculptural quality of the lawn space forms a calm green canvas. The lobby of the theatre looks out across an open space towards the main entrance to the old campus buildings. However, it is separated from the main buildings by an embankment almost six meters high and does not look directly towards the entrance.

Various text inscriptions add another level of meaning to the landscape. The client provided poems in various local languages and these were inscribed onto granite strips which were in turn worked into the paving in front of the theatre. The cornerstone of the building was built into the concrete bench.

设计说明

景观设计采用各种方式将艺术融入各种设计元素中，从诗歌艺术到草坪设计，草坪空间看起来如同一座干净的雕塑，又如一幅平静的绿色油画。从剧院的大厅往外，通过一个开放的空间，就能到达校园建筑的主要入口。然而，一个高6 m的路堤将它与主建筑分离开来，所以无法直接看到这个入口。

各种铭文雕刻又添加了另一个层面的景观意义。客户提供了当地不同方言写成的诗，这些诗文被刻到花岗岩带上，依次铺到剧院前。建筑的基石使用了混凝土材质。

| COMMERCIAL AND PUBLIC LANDSCAPE 商业公共景观 | CULTURAL SQUARE 文化广场 | URBAN SPACE 城市空间 |

Keywords 关键词

Natural Lighting 自然采光
Outdoor Space 户外空间
School Landscape 学校景观
Participation 可参与性

Location: Manassas Park, VA
Landscape Design: Siteworks

项目地点：弗吉尼亚州马纳萨斯公园
景观设计：Siteworks

Manassas Park Elementary School Landscape

马纳萨斯公园小学景观

Features 项目亮点

The design emphasizes the participation of the entire community, using the concept of sustainable development to create a space of discovery and imagination.

设计强调整个社区的参与性，运用可持续发展的理念营造了一个探索发现和想象的学校空间。

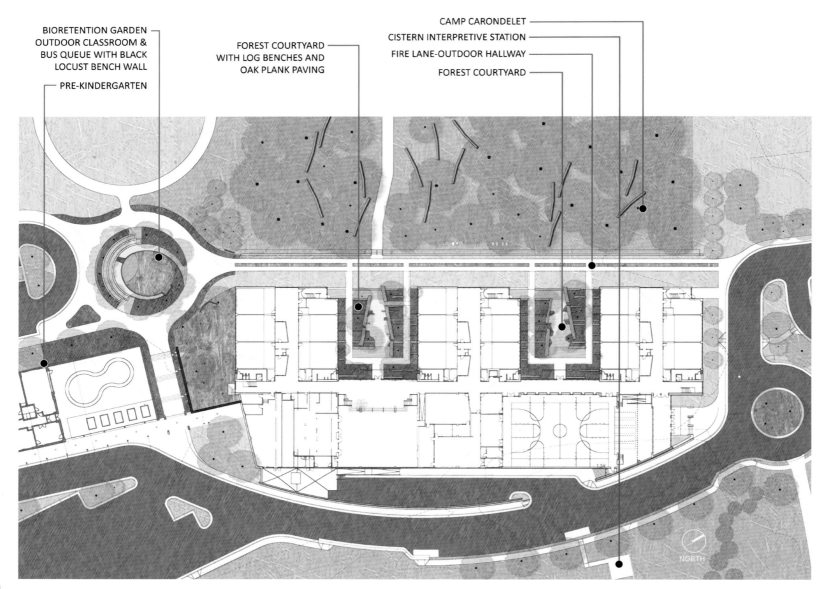

| COMMERCIAL SPACE 商业空间 | OFFICE SPACE 办公空间 | HEALTHCARE AND EDUCATION 医疗教育 |

COMMERCIAL AND PUBLIC LANDSCAPE 商业公共景观

CULTURAL SQUARE 文化广场

URBAN SPACE 城市空间

Overview

Studies have shown for years the benefits to student performance and health when schools are designed with fresh air, natural daylight and connections to the outdoors. MPES achieves those benefits while assuming the added responsibility of cultivating environmental stewards in their community of teachers, learners and parents.

项目概况

研究表明学校空气清新，自然采光好，并有户外活动空间，对学生健康成长有很大的帮助。该学校项目除具备这些条件外，还培养了教师、学生和家长爱护环境的意识。

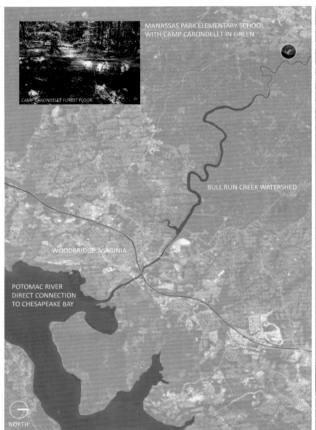
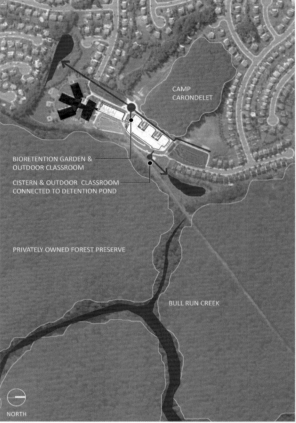

| COMMERCIAL SPACE 商业空间 | OFFICE SPACE 办公空间 | HEALTHCARE AND EDUCATION 医疗教育 |

Design Description

This LEED Gold certified project goes beyond sustainability checklists to create a school that challenges accepted paradigms in teaching and learning by actively involving the entire community in the design and ongoing operations of their school campus. MPES elicits a sense of imagination, discovery and wonder through the relationship it creates between a human community and the surrounding woodlands and watersheds within which it lives.

设计说明

该项目获得美国国家绿色建筑项目金奖,其设计大大超出之前对可持续发展的设计要求,在设计和实施过程中都积极调动了整个社区来参与。整个项目创造出一个社区和周围林地、水域相融合的环境,给人提供了一个探索发现和想象的空间。

COMMERCIAL AND PUBLIC LANDSCAPE 商业公共景观

CULTURAL SQUARE 文化广场

URBAN SPACE 城市空间

| COMMERCIAL AND PUBLIC LANDSCAPE 商业公共景观 | CULTURAL SQUARE 文化广场 | URBAN SPACE 城市空间 |

| COMMERCIAL SPACE 商业空间 | OFFICE SPACE 办公空间 | HEALTHCARE AND EDUCATION 医疗教育 |

| COMMERCIAL AND PUBLIC LANDSCAPE 商业公共景观 | CULTURAL SQUARE 文化广场 | URBAN SPACE 城市空间 |

Keywords 关键词

- Affiliated Green Land 附属绿地
- Site Landscape 场地景观
- Environment 环境
- Space 空间

Location: Tempe, AZ, USA
Client: Arizona State University
Landscape Design: Ten Eyck Landscape Architects

项目地点：美国亚利桑那州立大学坦佩校区
客　　户：亚利桑那州立大学
景观设计：Ten Eyck 景观设计工作室

The Biodesign Institute at Arizona State University

亚利桑那州立大学生物设计研究所

Features 项目亮点

Drawing the site green land layout based on the master plan development, the design creates a close, varied landscape space.

设计从整体性的开发要求着手，根据场地的环境对附属绿地进行规划布局，营造了一个联系密切、形态多样的景观空间。

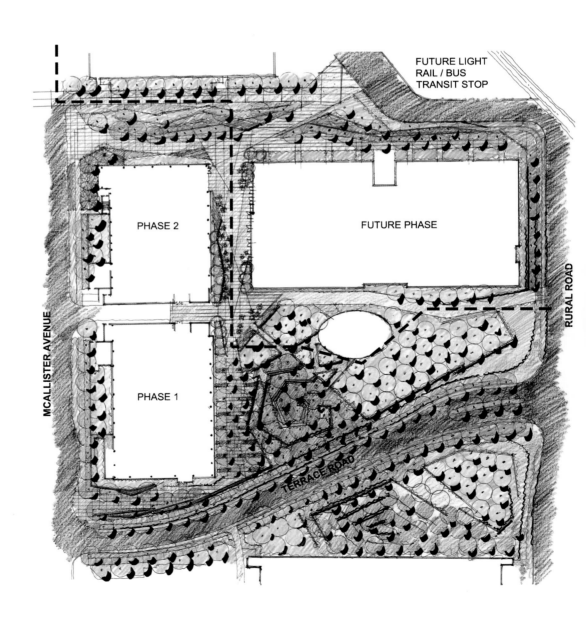

| COMMERCIAL SPACE 商业空间 | OFFICE SPACE 办公空间 | HEALTHCARE AND EDUCATION 医疗教育 |

Overview

The 4-acre Biodesign Institute site is located at the Tempe campus of Arizona State University. It is east of Rural Road—a major arterial connecting Tempe with Scottsdale. Terrace Road enters the site from Rural Road and is a major entry point to the east campus zone. The south boundary of the project originally was Terrace Road but during the master planning phase an existing sterile retention basin next to a major parking garage south of Terrace was included by the urging of the landscape designer. The site is bounded by Tyler to the north and McAllister to the west.

项目概况

该项目占地 4 英亩（约 16 187 m²），是位于亚利桑那州立大学（ASU）坦佩校区东部的生物设计研究所的附属绿地。场地东侧为连接坦佩与斯科茨代尔两市的主干道田园路，西侧为 McAllister 大街，北侧为泰勒大街，南侧原为台地路，但总体规划阶段景观设计师要求将路南大型车库旁的一个不毛的滞留池也纳入设计范围，台地路由此变成了横穿场地的路，同时它也是进入校园东区的一个主要通道。

COMMERCIAL AND PUBLIC LANDSCAPE 商业公共景观

CULTURAL SQUARE 文化广场

URBAN SPACE 城市空间

| COMMERCIAL SPACE 商业空间 | OFFICE SPACE 办公空间 | HEALTHCARE AND EDUCATION 医疗教育 |

COMMERCIAL AND PUBLIC LANDSCAPE 商业公共景观

| CULTURAL SQUARE 文化广场 | URBAN SPACE 城市空间 |

| COMMERCIAL SPACE 商业空间 | OFFICE SPACE 办公空间 | HEALTHCARE AND EDUCATION 医疗教育 |

COMMERCIAL AND PUBLIC LANDSCAPE 商业公共景观

CULTURAL SQUARE 文化广场

URBAN SPACE 城市空间

Keywords 关键词

Green Roof 绿色屋顶
Vegetation Design 植被设计
Natural Environment 自然环境
Sustainability 可持续性

Location: San Pedro Garza García, Nuevo León, México
Landscape Design: HARARI Landscape Architecture
Leader: Claudia Harari
Team: Silverio Sierra, Mario Cavazos, Ernesto Marroquín, Susana García
Architectural Design: Legorreta + Legorreta
Total area of concrete slabs: 13,807 m²
Total area of green roofs: 9,500 m² (69% of the slabs total)
Minimal thickness of extensive green roofs: 210 mm
Minimal thickness of intensive green roofs: 600 mm

项目地点：墨西哥新莱昂州 San Pedro Garza García
景观设计：HARARI 景观设计事务所
项目负责人：Claudia Harari
设计团队：Silverio Sierra, Mario Cavazos, Ernesto Marroquín, Susana García
建筑设计：Legorreta + Legorreta
水泥场总面积：13,807 m²
屋顶绿化总面积：9,500 m²
粗放型绿色屋顶最小厚度：210 mm
精细型绿色屋顶最小厚度：600 mm

Zambrano Hellion Hospital

Zambrano Hellion 医院

Features 项目亮点

With roof green and vegetal selection, the landscape design evolves from the architecture forms, the immediate natural environment and the visual towards the distant mountainous landscape.

该项目通过空中绿化、植被选择等景观设计措施，将建筑形态、自然环境及远处的山地景观巧妙地融合在了一起。

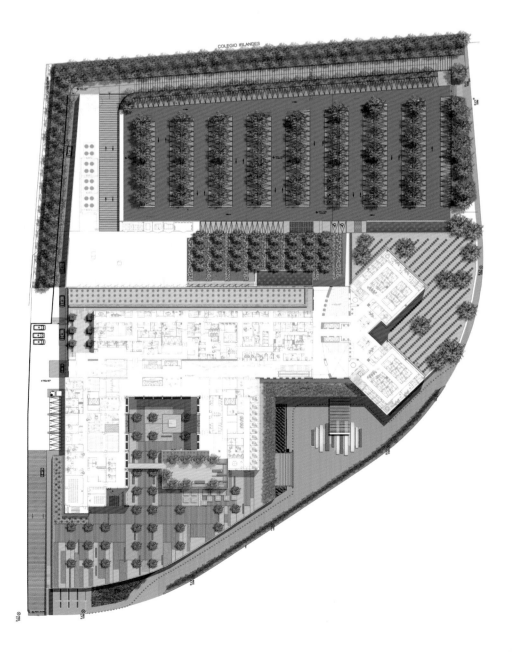

| COMMERCIAL SPACE 商业空间 | OFFICE SPACE 办公空间 | HEALTHCARE AND EDUCATION 医疗教育 |

Overview

The green areas system is a key part of the building's infrastructure. It develops entirely on concrete slabs which are the parking and the hospital areas. Approximately 9,000 m² of membranes and special substrates take part in one of the largest green roofs in Latin America. The landscape design evolves from the architecture forms, the immediate natural environment and the visual towards the distant mountainous landscape.

项目概况

景观绿化是该医院设施的重要组成部分。整座医院及其停车场都位于水泥场地上，约9 000 m²的空中绿化使之成为拉丁美洲最大的绿色屋顶之一。通过景观设计，该项目成功地将建筑形态、自然环境及远处的山地景观融合在了一起。

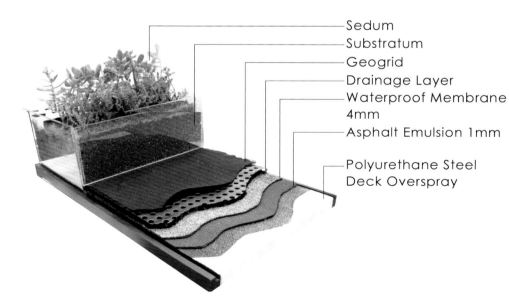

Green Roof Detail

| COMMERCIAL AND PUBLIC LANDSCAPE 商业公共景观 | CULTURAL SQUARE 文化广场 | URBAN SPACE 城市空间 |

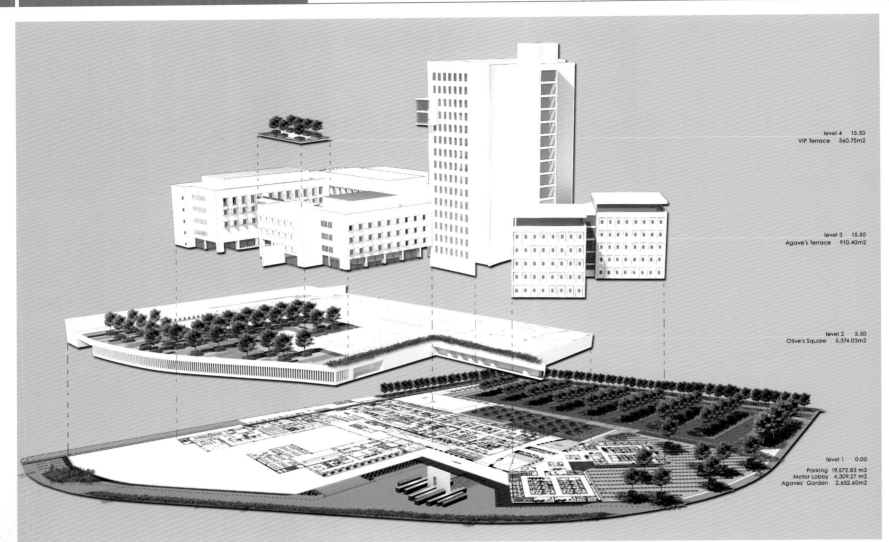

level 4 15.50
VIP Terrace 560.75m2

level 3 15.50
Agave's Terrace 910.40m2

level 2 5.50
Olive's Square 5,374.03m2

level 1 0.00
Parking 19,572.83 m2
Motor Lobby 4,309.27 m2
Agaves' Garden 2,652.60m2

| COMMERCIAL SPACE 商业空间 | OFFICE SPACE 办公空间 | HEALTHCARE AND EDUCATION 医疗教育 |

Design Description

Vegetal Selection—The natural selection is based in two aspects:

1. Sustainability: native and adapted species with low water consumption and maintenance; and high resistance to extreme temperatures and wind.

2. Symbolism: olives and oaks as symbols of life, longevity and hope. Species of aromatic plants related to the processes of relaxation, natural medicine and healing.

The intervened areas are distributed in four levels:

Level 0.00 – Vehicular/Pedestrian Access and Parking

Level +5.50 – Olive Terrace

Level +15.50 – Maternity Area

Level +25.50 – VIP Terrace of Suites

The slab's temperature reduction coefficient is of 25 degrees below the green layer.

设计说明

植被设计——植物品种的选择主要基于两个方面的考虑：

1. 可持续性：设计倾向于选择耐旱性强、易维护且能适应极端气候的本土植物和引进植物。

2. 象征性：橄榄树和橡树象征着生命、长寿和希望。而芬芳的植物则是天然的"良药"，有益于放松，有助于康复。

景观规划区域主要分布于四层：

地面层设置车行道、人行道及停车场；

5.5 m 高的位置设有橄榄露台；

15.5 m 处是产科区域；

25.5 m 处是 VIP 套房露台。

水泥板的温度低于绿化保温层的25°。

COMMERCIAL AND PUBLIC LANDSCAPE 商业公共景观

| CULTURAL SQUARE 文化广场 | URBAN SPACE 城市空间 |

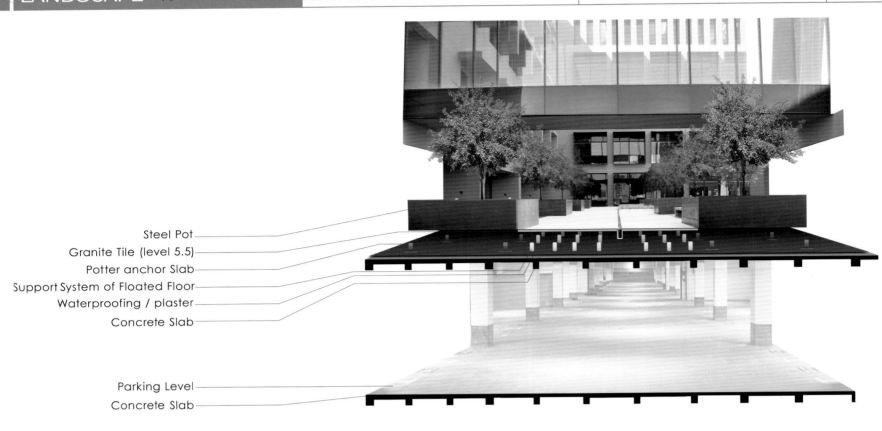

- Steel Pot
- Granite Tile (level 5.5)
- Potter anchor Slab
- Support System of Floated Floor
- Waterproofing / plaster
- Concrete Slab

- Parking Level
- Concrete Slab

| COMMERCIAL SPACE 商业空间 | OFFICE SPACE 办公空间 | HEALTHCARE AND EDUCATION 医疗教育 |

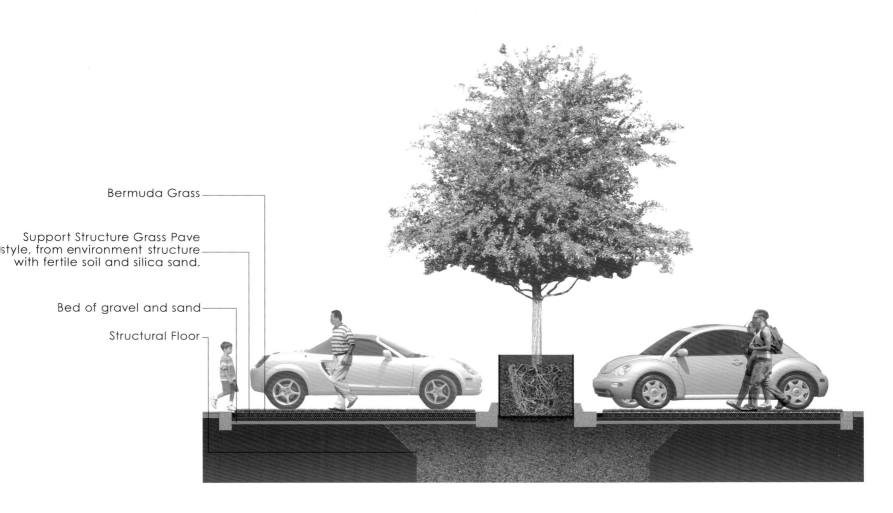

- Bermuda Grass
- Support Structure Grass Pave style, from environment structure with fertile soil and silica sand.
- Bed of gravel and sand
- Structural Floor

- Agave Tequilana
- Gravel
- Lightening
- Geotextile Filter Layer 250gr. 2mm
- Modular Drain Prefabricated
- Prefabricated Membrane with polyester reinforcements and non-woven agent 4mm
- Layer of prefabricated Membrane with polyester reinforcements 3mm
- Asphalt Solvent 1mm
- Concrete Slab

Agaves' Garden Detail

COMMERCIAL AND PUBLIC LANDSCAPE 商业公共景观 | CULTURAL SQUARE 文化广场 | URBAN SPACE 城市空间

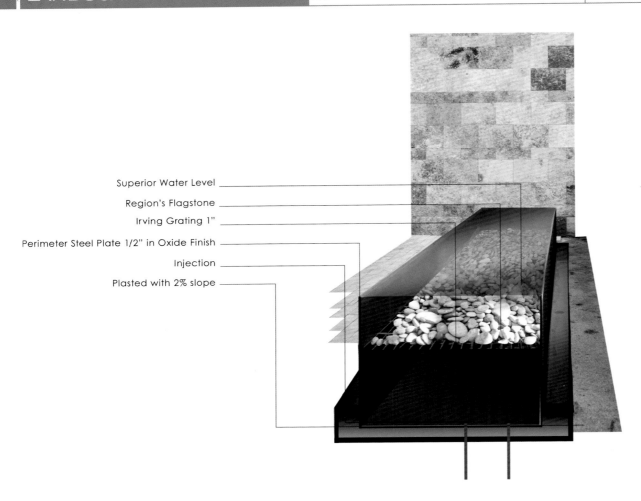

- Superior Water Level
- Region's Flagstone
- Irving Grating 1"
- Perimeter Steel Plate 1/2" in Oxide Finish
- Injection
- Plasted with 2% slope

| COMMERCIAL SPACE 商业空间 | OFFICE SPACE 办公空间 | HEALTHCARE AND EDUCATION 医疗教育 |

COMMERCIAL AND PUBLIC LANDSCAPE 商业公共景观

CULTURAL SQUARE 文化广场

URBAN SPACE 城市空间

| COMMERCIAL SPACE 商业空间 | OFFICE SPACE 办公空间 | HEALTHCARE AND EDUCATION 医疗教育 |